The Boundaries of
Natural Science

The Boundaries of Natural Science

Eight lectures given in Dornach, Switzerland
September 27–October 3, 1920

by
Rudolf Steiner

with a foreword
by
Saul Bellow

ANTHROPOSOPHIC
PRESS

Translated by Frederick Amrine and Konrad Oberhuber from shorthand reports unrevised by the lecturer, from the 4th edition (1969) of the German text published under the title *Grenzen der Naturerkenntnis* (Vol. 322 in the Bibliographic Survey). The translation was supported in part by a grant from the *Austro-American Friendship Association of Boston.*.

Gratitude is expressed for permission to quote in the Introduction from *Collected Works of Paul Valéry*, ed. Jackson Matthews, Bollingen Series XLV, Vol. 9: *Masters and Friends*, trans. Martin Turnell © 1968 Princeton University Press.

Library of Congress Cataloging in Publication Data

Steiner, Rudolf, 1861-1925.
  The boundaries of natural science.

  "Translated from shorthand reports unrevised by the lecturer, from the 4th edition (1969) of the German text published under the title Grenzen der Naturerkenntnis (Vol. 322 in the Bibliographic survey)"—T.p. verso.
  1. Anthroposophy—Addresses, essays, lectures.
  2. Science—Methodology-Addresses, essays, lectures.
  I. Title.
BP595.S85372513 1982   299'.935   83-9943
ISBN 978-0-88010-187-5

Cover design by Peter Stebbing

Table of Contents

Foreword

The audience attending this series of lectures in 1920 was at once informed by Steiner that he proposed to consider the connections between natural science and social renewal.

Everyone agrees, he says, that such a renewal requires a renewal of our thinking (one must remember that he was speaking of the groping and soul-searching that followed the great and terrible war of 1914-18), yet not everyone "imagines something clear and distinct when speaking in this way."

Steiner then sketches rapidly the effects of the scientific world-view on the modern social order. Scientific progress has made us very confident of our analytical powers. Inanimate nature, we are educated to believe, will eventually become transparently intelligible. It will yield all its secrets under scientific examination, and we will be able to describe it with mathematical lucidity. After we have conquered the inorganic we will proceed to master the organic world by the same means.

The path of scientific progress however has not been uniformly smooth. Steiner reminds us that by the end of the 19th century doubts concerning the origins of scientific knowledge had arisen within the scientific community itself, and in a famous and controversial lecture the physiologist Du Bois Reymond asked the question, How does consciousness arise out of material processes? What is the source of the consciousness with which we examine the outer world? To this Du Bois Reymond answers, *Ignorabimus*—we shall never know.

In this *Ignorabimus* Steiner finds a parallel to an earlier development, that of medieval Scholasticism. Scholastic thinking had made its way to the limits of the supersensible world. Modern natural science has also reached a limit. This limit is delineated by two concepts: "matter"—which is everywhere assumed to be within the sensory realm but nowhere actually to be found—and consciousness, which is assumed to originate within the same world, "although no one can comprehend how." Can we fathom the fact of consciousness with explanations conceived in observing external nature? Steiner argues that we cannot. He suggests that scientific research is entangling itself in a web, and that only outside this web can we find the real world. The great victories of science have subdued our minds. We accept the all pervading scientific method. It has transformed the earth. Nevertheless it seems incapable of understanding its own deepest sources. Scientific method as we of the modern world define it can bring us only to the *Ignorabimus* because it is powerless to explain the consciousness that directs it. In our study of nature, and by means of our concept of matter, we have made everything very clear, but this clarity does not give us Man. Him we have lost. And the lucidity to which we owe our great successes in the study of the external world is rejected by consciousness itself. For in the depths of consciousness there lies a will, and this will revolts when lucid science tries to "think" Man as it thinks external nature.

To conclude from this that Steiner is "anti-science" would be a great mistake. To him science is a necessary, indeed indispensable stage in the development of the human spirit. The scientific examination of the external world awakens consciousness to clear concepts and it is by means of clear conceptual thinking that we become fully human. Spiritual development requires a full understanding of pure thought, and pure thought is thought devoid of sensory impressions. "Countless philosophers have expounded the view that pure thinking does not exist, but is bound to contain traces, however diluted, of sense perception. A strong impression is left that philosophers who maintain this have never really studied mathematics, or gone into the difference between analytical and empirical physics," Steiner writes. Mathematical thought is thought detached from the sense world, and as it is entirely based upon rules of reason that are universal it offers spiritual communion to mankind, as well as a union with reality. It is moreover a *free* activity. Spiritual training, says Steiner, reveals it to be not only sense-free but also brain-free. The operations of thought are directed by spiritual powers. Pure thinking leads to the discovery of freedom and leads us to the realm of spirit. And Steiner tells us explicitly that out of sense-free thinking "there can flow impulses to moral action. . . . One experiences pure spirit by observing, by actually observing how moral forces flow into sense-free thinking." This is something very different from mystical experience, for it is a result of spiritual training, of a sort of scientific discipline through which we discover more organs of knowledge than are available to those who limit themselves, as modern philosophers do, to scientific orthodoxy and to ordinary consciousness. In the last lecture of the present series Steiner speaks of advanced forms of consciousness, of a more acute inner activity, and of higher forms of knowledge.

Contemporary thinkers are often strongly attracted to these higher forms. They approach them enthusiastically, frequently write of them vividly but in the end reject them as retrograde or atavistic, unworthy of a fully accredited modern philosopher.

Paul Valéry, a poet who devoted years of his life to the study of mathematics and who wrote interestingly on Descartes and Pascal, provides us with an excellent example of this in his *Address in Honor of Goethe*. Goethe fascinates Valéry, for Goethe too was a poet who found it necessary to go beyond poetry—"the great apologist of the world of Appearances," Valéry calls him. He says, "I sometimes think that there exists for some people, as there existed for him, an *external life* which has an intensity and a depth at least equal to the intensity and depth that we ascribe to the inner darkness and the mysterious discoveries of the ascetics and the Sufis." Goethe is an investigative and not merely a reactive poet. Valéry greatly admires his botanical work, seeing in it one of "the profound nodal points of his great mind." He goes on to say, "this desire to trace in living things a will to metamorphosis may have been derived from his early contact with certain doctrines, half poetic, half esoteric, which were highly esteemed by the ancients and which, at the end of the eighteenth century, initiates took to cultivating again. The rather seductive if extremely imprecise idea of Orphism, the magical idea of assuming the existence of some unknown hidden principle of life, some tendency towards a higher form of life in every animate and inanimate thing; the idea that a spirit was fermenting in every particle of reality and that it was therefore not impossible to work by the ways of the spirit on everything and every being insofar as it contains a spirit, is among the ideas which bear witness to the persistence of a kind of primitive reasoning and at the same time of an impulse which of its nature generates poetry or personification. Goethe appears to have been deeply imbued with the feeling of this power, which satisfied the poet in him and stimulated the naturalist."

What Valéry assumes here is that there is only one single legitimate method of examining natural phenomena. As a poet he sympathizes with imaginative knowledge, as a thinker he strikes a note of regret and even condolence. "It is one of the clearest examples of transition from poetic thought to scientific theory, or of a fact brought to light by way of a harmony discovered by intuition. Observation verifies what the inner artist has divined. . . . But his great gift of analogy came into conflict with his logical faculties." And the logical faculties, strictly circumscribed, must be obeyed. Magic and primitive reasoning, alas, will not do says the analytical intellect of Valéry.

Steiner had devoted many years of study to Goethe. He was the editor of Goethe's scientific works and in his lectures often refers to him. And there is no nostalgia for "Orphism" in Steiner, no "magic" or "primitive reasoning." He too is a modern thinker. What distinguishes him from most others is his refusal to stop at what he calls "the boundary of the material world." And how does one pass beyond this boundary? By a discipline that takes us from ordinary consciousness and familiarizes us with consciousness of another kind, by finding the path that leads us into Imagination. "It is possible to pursue this path in a way consonant with Western life," he writes, "if we attempt to surrender ourselves completely to the world of outer phenomena, so that we allow them to work upon us without thinking about them, but still perceiving them. In ordinary waking life, you will agree, we are constantly perceiving, but actually in the very process of doing so we are continually saturating our percepts with concepts; in scientific thinking we interweave percepts and concepts entirely systematically, building up systems of concepts. . . . One can become capable of such acute inner activity that one can exclude and suppress conceptual thinking from the process of perception and surrender oneself to bare percepts." This is not a depreciation of thought. Rather, it releases the imagination. One "acquires a potent psychic force 'when one is able' to absorb the external world free from concepts." Steiner says, "Man is given over to the external world continually, from birth onwards. Nowadays this giving-over of oneself to the external world is held to be nothing but abstract perception or abstract cognition. This is not so. We are surrounded by a world of color, sound and warmth and by all kinds of sensory impressions." The cosmos communicates with us also through color, sound and warmth. "Warmth is something other than warmth; light something other than light in the physical sense; sound is something other than physical sound. Through our sensory impressions we are conscious only of what I would term external sound and external color. And when we surrender ourselves to nature we do not encounter the ether-waves, atoms and so on of which modern physics and physiology dream; rather, it is spiritual forces that are at work, forces that fashion us between birth and death into what we are as human beings." I have thought it best not to interpose myself but to allow Steiner to speak for himself, for he is more than a thinker, he is an initiate and only he is able to communicate what he has experienced. The human mind, he tells us, must learn to will pure thinking, but it must learn also how to set conceptual thinking aside and to live within the phenomena. "It is through phenomenology, and not abstract metaphysics, that we attain knowledge of the spirit by consciously observing, by raising to consciousness, what we would otherwise do unconsciously; by observing how through the sense-world spiritual forces enter into our being and work formatively upon it."

We cannot even begin to think of social renewal until we

have considered these questions. What is reality in the civilized West? "A world of outsides without insides," says Owen Barfield, one of the best interpreters of Steiner. A world of quantities without qualities, of souls devoid of mobility and of communities which are more dead than alive.

Saul Bellow

I

The theme of this cycle of lectures was not chosen because it is traditional within academic or philosophical disciplines, as though we thought epistemology or the like should appear within our courses. Rather, it was chosen as the result of what I believe to be an open-minded consideration of the needs and demands of our time. The further evolution of humanity demands new concepts, new notions, and new impulses for social life generally: we need ideas which, when realized, can create social conditions offering to human beings of all stations and classes an existence that seems to them humane. Already, to be sure, it is being said in the widest circles that social renewal must begin with a renewal of our thinking.[1] Yet not everyone in these widest circles imagines something clear and distinct when speaking in this way. One does not ask: whence shall come the ideas upon which one might found a social economy offering man a humane existence? That portion of humanity which has received an education in the last three to four centuries, but particularly since the nineteenth century, has been raised with certain ideas that are outgrowths of the scientific world view or entirely schooled in it. This is particularly true of those who have undergone some academic training. Only those working in fields other than the sciences believe that natural science has had little influence on their pursuits. Yet it is easy to demonstrate that even in the newer, more progressive theology, in history and in jurisprudence—every-

1

where can be found scientific concepts such as those that arose from the scientific experiments of the last centuries, so that traditional concepts have in a certain way been altered to conform to the new. One need only allow the progress of the new theological developments in the nineteenth century to pass before the mind's eye. One sees, for example, how Protestant theology has arrived at its views concerning the man, Jesus, and the nature of Christ, because at every turn it had in mind certain scientific conceptions that it wanted to satisfy, against which it did not want to sin. At the same time, the old, instinctive ties within the social order began to slacken: they gradually ceased to hold human life together. In the course of the nineteenth century it became increasingly necessary to replace the instincts according to which one class subordinated itself to another, the instincts out of which the new parliamentary institutions, with all their consequences, have come with more-or-less conscious concepts. Not only in Marxism but in many other movements as well there has come about what one might call a transformation of the old social instincts into conscious concepts.

But what was this new element that had entered into social science, into this favorite son of modern thought? It was the conceptions, the new mode of thinking that had been developed in the pursuit of natural science. And today we are faced with the important question: how far shall we be able to progress within a web of social forces woven from such concepts? If we listen to the world's rumbling, if we consider all the hopeless prospects that result from the attempts that are made on the basis of these conceptions, we are confronted with a dismal picture indeed. One is then faced with the portentous question: how does it stand with those very concepts that we have acquired from natural science and now wish to apply to our lives, concepts that—this has become clearly evident in many areas already—are actually rejected by life itself? This vital question, this burning question with which our age confronts us, was the occasion of my choosing the theme, "The Boundaries of Natural Science." Just this question requires that I treat the theme in such a way that we receive an overview of what natural science can and cannot contribute to an appropriate social order and an idea of the kind of scientific research, the kind of world view to which one would have to turn in order to confront seriously the demands made upon us by our time.

What is it we see if we consider the method according to which one thinks in scientific circles and how others have been influenced in their thinking by those circles? What do we see? We see first of all that an attempt is made to acquire data and to order it in a lucid system with the help of clear concepts. We see how an attempt is made to order the data gathered from inanimate nature by means of the various sciences—mechanics, physics, chemistry, etc.—to order them in a systematic manner but also to permeate the data with certain concepts so that they become intelligible. With regard to inanimate nature, one strives for the greatest possible clarity, for crystal-clear concepts. And a consequence of this striving for lucid concepts is that one seeks, if it is at all possible, to permeate everything that one finds in one's environment with mathematical formulae. One wants to translate data gathered from nature into clear mathematical formulae, into the transparent language of mathematics.

In the last third of the nineteenth century, scientists already believed themselves very close to being able to give a mathematical-mechanical explanation of natural phenomena that would be thoroughly transparent. It remained for them only to explain the little matter of the atom. They wanted to reduce it to a point-force [*Kraftpunkt*] in order to be able to express its position and momenta in mathematical formulae. They believed they would then be justified in say-

3

ing: I contemplate nature, and what I contemplate there is in reality a network of interrelated forces and movements wholly intelligible in terms of mathematics. Hence there arose the ideal of the so-called "astronomical explanation of nature," which states in essence: just as one brings to expression the relationships between the various heavenly bodies in mathematical formulae, so too should one be able to compute everything within this smallest realm, within the "little cosmos" of atoms and molecules, in terms of lucid mathematics. This was a striving that climaxed in the last third of the nineteenth century: it is now on the decline again. Over against this striving for a crystal-clear, mathematical view of the world, however, there stands something entirely different, something that is called forth the moment one tries to extend this striving into realms other than that of inanimate nature. You know that in the course of the nineteenth century the attempt was made to carry this point of view, at least to some extent, into the life sciences. And though Kant had said that a second Newton would never be found who could explain living organisms according to a causal principle similar to that used to explain inorganic nature, Haeckel could nevertheless claim that this second Newton had been found in Darwin, that Darwin had actually tried, by means of the principle of natural selection, to explain how organisms evolve in the same "transparent" terms. And one began to aim for just such a clarity, a clarity at least approaching that of mathematics, in all explanations, proceeding all the way up to the explanation of man himself. Something thereby was fulfilled which certain scientists explained by saying that man's need to understand the causes of phenomena is satisfied only when he arrives at such a transparent, lucid view of the world.

And yet over against this there stands something entirely different. One comes to see that theory upon theory has

4

been contrived in order to construct a view of the world such as I have just described, and ever and again those who strove for such a view of the world called forth—often immediately—their own opposition. There always arose the other party, which demonstrated that such a view of the world could never produce valid explanations, that such a view of the world could never ultimately satisfy man's need to know. On the one hand it was argued how necessary it is to keep one's world view within the lucid realm of mathematics, while on the other hand it was shown that such a world view would, for example, remain entirely incapable of constructing even the simplest living organism in thought of mathematical clarity or, indeed, even of constructing a comprehensible model of organic substance. It was as though the one party continually wove a tissue of ideas in order to explain nature, and the other party—sometimes the same party—continually unraveled it.

It has been possible to follow this spectacle—for it seems just that to anyone who is able to view it with an unprejudiced eye—within the scientific work and striving of the last fifty years especially. If one has sensed the full gravity of the situation, that with regard to this important question nothing but a weaving and unraveling of theories has taken place, one can pose the question: is not the continual striving for such a conceptual explanation of phenomena perhaps superfluous? Is not the proper answer to any question that arises when one confronts phenomena perhaps that one should simply allow the facts to speak for themselves, that one should describe what occurs in nature and forgo any more detailed accounting? Is it not possible that all such explanations show only that humanity is still tied to its mother's apron strings, that humanity in its infancy sought a kind of luxury? Would not humanity, come of age, have to say to itself: we must not strive at all for such explanations;

5

we get nowhere in that way and must simply extirpate the need to know? Why not? As we become older we outgrow the need to play; why, if we were justified in doing so, should we not simply outgrow the need for explanations?

Just such a question could already emerge in the most extraordinarily significant way when, on August 14, 1872, Du Bois-Reymond stood before the Second General Meeting of the Association of German Scientists and Physicians to deliver his famous address, "The Boundaries of Natural Science" ["*Grenzen des Naturerkennens*"], an address still worthy of consideration today. Yet despite the amount that has been written about this address by the important physiologist, Du Bois-Reymond, many still do not realize that it represents one of the important junctures in the evolution of the modern world view.

In medieval Scholasticism all of man's thinking, all of his notional activity, was determined by the view that one could explain the broad realms of nature in terms of certain concepts but that one had to draw the line upon reaching the supersensible. The supersensible had to be the object of revelation. They felt that man should stand in a relation to the supersensible in such a way that he would not even wish to penetrate it with the same concepts he formed concerning the realms of nature and external human existence. A limit was set to knowledge on the side of the supersensible, and it was strongly emphasized that such a limit had to exist, that it simply lay within human nature and the order of the universe that such a limit be recognized. This placement of a limit to knowledge was then renewed from an entirely different side by thinkers and researchers such as Du Bois-Reymond. They were no longer Schoolmen, no longer theologians, but just as the medieval theologian, proceeding according to his own mode of thinking, had set a limit to knowledge at the supersensible, so these thinkers and researchers set a limit at the sensible. The limit was meant to apply above all to the realm of external sensory data.

There were two concepts in particular that Du Bois-Reymond had in mind, which to him established the limits natural science could reach but beyond which it could not proceed. Later he increased that number by five in his lecture, "The Seven Enigmas of the World," but in the first lecture he spoke of the two concepts, "matter" and "consciousness." He said that when contemplating nature we are forced, in thinking systematically, to apply concepts in such a way that we eventually arrive at the notion of matter. Just what this mysterious entity in space we call "matter" is, however, we can never in any way resolve. We must simply assume the concept "matter," though it is opaque. If only we assume this opaque concept "matter," we can apply our mathematical formulae and calculate the movements of matter in terms of the formulae. The realm of natural phenomena becomes comprehensible if only we can posit this "opaque" little point millions upon millions of times. Yet surely we must also assume that it is this same material world that first builds up our bodies and unfolds its own activity within them, so that there rises up within us, by virtue of this corporeal activity, what eventually becomes sensation and consciousness. On the one hand we confront a world of natural phenomena requiring that we construct a concept of "matter," while on the other hand we confront ourselves, experience the fact of consciousness, observe its phenomena, and surmise that whatever it is we assume to be matter must also lie at the foundation of consciousness. But how, out of these movements of matter, out of inanimate, dead movement, there arises consciousness, or even simple sensation, is a mystery that we cannot possibly fathom. This

7

is the other pole of all the uncertainties, all the limits to knowledge: how can we explain consciousness, or even the simplest sensation?

With regard to these two questions, then—What is matter? How does consciousness arise out of material processes? —Du Bois-Reymond maintains that as researchers we must confess: *ignorabimus*, we shall never know. That is the modern counterpart to medieval Scholasticism. Medieval Scholasticism stood at the limit of the supersensible world. Modern natural science stands at the limit delineated in essence by two concepts: "matter," which is everywhere assumed within the sensory realm but nowhere to be found, and "consciousness," which is assumed to originate within the sense world, although one can never comprehend how.

If one considers this development of modern scientific thought, must one not then say to oneself that scientific research is entangling itself in a kind of web, and only outside of this web can one find the world? For in the final analysis it is there, where matter haunts space, that the external world lies. If this is the one place into which one cannot penetrate, one has no way in which to come to terms with life. Within man one finds the fact of consciousness. Does one come at all near to it with explanations conceived in observing external nature? If in one's search for explanations one pulls up short at human life, how, then, can one arrive at notions of how to live in a way worthy of a human being? How, if one cannot understand the existence or the essence of man according to the assumptions one makes concerning that existence?

As this course of lectures progresses it shall, I believe, become evident beyond any doubt that it is the impotence of the modern scientific method that has made us so impotent in our thinking about social questions. Many today still do not perceive what an important and essential connection exists between the two. Many today still do not perceive that when in Leipzig on August 14, 1872 Du Bois-Reymond spoke his *ignorabimus*, this same *ignorabimus* was spoken also with regard to all social thought. What this *ignorabimus* actually meant was: we stand helpless in the face of real life; we have only shadowy concepts; we have no concepts with which to grasp reality. And now, almost fifty years later, the world demands just such concepts of us. We must have them. Such concepts, such impulses, cannot come out of lecture-halls still laboring in the shadow of this *ignorabimus*. That is the great tragedy of our time. Here lie questions that must be answered.

We want to proceed from fundamental principles to such an answer and above all to consider the question: is there not perhaps something more intelligent that we as human beings could do than what we have done for the last fifty years, namely tried to explain nature after the fashion of ancient Penelope, by weaving theories with one hand and unraveling them with the other? Ah yes, if only we could, if only we could stand before nature entirely without thoughts! But we cannot: to the extent that we are human beings and wish to remain human beings we cannot. If we wish to comprehend nature, we must permeate it with concepts and ideas. Why must we do that?

We must do that, ladies and gentlemen, because only thereby does consciousness awake, because only thereby do we become conscious human beings. Just as each morning upon opening our eyes we achieve consciousness in our interaction with the external world, so essentially did consciousness awake within the evolution of humanity. Consciousness, as it is now, was first kindled through the interaction of the senses and thinking with the outer world. We can watch the historical development of consciousness in the interaction of man's senses with outer nature. In this

9

process consciousness gradually was kindled out of the dull, sleepy cultural life of primordial times. Yet one must only consider with an open mind this fact of consciousness, this interaction between the senses and nature, in order to observe something extraordinary transpiring within man. We must look into our soul to see what is there, either by remaining awhile before fully awakening within that dull and dreamy consciousness or by looking back into the almost dreamlike consciousness of primordial times. If we look within our soul at what lies submerged beneath the surface consciousness arising in the interaction between senses and the outer world, we find a world of representations, faint, diluted to dream-pictures with hazy contours, each image fading into the other. Unprejudiced observation establishes this. The faintness of the representations, the haziness of the contours, the fading of one representation into another: none of this can cease unless we awake to a full interaction with external nature. In order to come to this awakening—which is tantamount to becoming fully human—our senses must awake every morning to contact with nature. It was also necessary, however, for humanity as a whole to awake out of a dull, dreamlike vision of primordial worlds within the soul to achieve the present clear representations.

In this way we achieve the clarity of representation and the sharply delineated concepts that we need in order to remain awake, to remain aware of our environment with a waking soul. We need all this in order to remain human in the fullest sense of the word. But we cannot simply conjure it all up out of ourselves. We achieve it only when our senses come into contact with nature: only then do we achieve clear, sharply delineated concepts. We thereby develop something that man must develop for his own sake—otherwise consciousness would not awake. It is thus not an abstract "need for explanations," not what Du Bois-

Reymond and other men like him call "the need to know the causes of things," that drives us to seek explanations but the need to become human in the fullest sense through observing nature. We thus may not say that we can outgrow the need to explain like any other child's play, for that would mean that we would not want to become human in the fullest sense of the word—that is to say, not want to awake in the way we must awake.

Something else happens in this process, however. In coming to such concepts as we achieve in contemplating nature, we at the same time impoverish our inner conceptual life. Our concepts become clear, but their compass becomes diminished, and if we consider exactly what it is we have achieved by means of these concepts, we see that it is an external, mathematical-mechanical lucidity. Within that lucidity, however, we find nothing that allows us to comprehend life. We have, as it were, stepped out into the light but lost the very ground beneath our feet. We find no concepts that allow us to typify life, or even consciousness, in any way. In exchange for the clarity we must seek for the sake of our humanity, we have lost the content of that for which we have striven. And then we contemplate nature around us with our concepts. We formulate such complex ideas as the theory of evolution and the like. We strive for clarity. Out of this clarity we formulate a world view, but within this world view it is impossible to find ourselves, to find man. With our concepts we have moved out to the surface, where we come into contact with nature. We have achieved clarity, but along the way we have lost man. We move through nature, apply a mathematical-mechanical explanation, apply the theory of evolution, formulate all kinds of biological laws; we explain nature; we formulate a view of nature—within which man cannot be found. The abundance of content that we once had has been lost, and we are

11

confronted with a concept that can be formed only with the clearest but at the same time most desiccated and lifeless thinking: the concept of matter. And an *ignorabimus* in the face of the concept of matter is essentially the confession: I have achieved clarity; I have struggled through to an awakening of full consciousness, but thereby I have lost the essence of man in my thinking, in my explanations, in my comprehension.

And now we turn to look within. We turn away from matter to consider the inner realm of consciousness. We see how within this inner realm of consciousness representations pass in review, feelings come and go, impulses of will flash through us. We observe all this and notice that when we attempt to bring the inner realm into the same kind of focus that we achieved with regard to the external world, it is impossible. We seem to swim in an element that we cannot bring into sharp contours, that continually fades in and out of focus. The clarity for which we strive with regard to outer nature simply cannot be achieved within. In the most recent attempts to understand this inner realm, in the Anglo-American psychology of association, we see how, following the example of Hume, Mill, James, and others, the attempt was made to impose the clarity attained in observation of external nature upon inner sensations and feelings. One attempts to impose clarity upon sensation, and this is impossible. It is as though one wanted to apply the laws of flight to swimming. One does not come to terms at all with the element within which one has to move. The psychology of association never achieves sharpness of contour or clarity regarding the phenomenon of consciousness. And even if one attempts with a certain sobriety, as Herbart has done, to apply mathematical computation to human mental activity [*das Vorstellen*], to the human soul, one finds it possible, but

12

the computations hover in the air. There is no place to gain a foothold, because the mathematical formulae simply cannot comprehend what is actually occurring within the soul. While one loses man in coming to clarity regarding the external world, one finds man, to be sure—it goes without saying that one finds man when one delves into consciousness—but there is no hope of achieving clarity, for one swims about, borne hither and thither in an insubstantial realm. One finds man, but one cannot find a valid image of man.

It was this that Du Bois-Reymond felt very clearly but was able to express only much less clearly—only as a kind of vague feeling about scientific research on the whole—when in August 1872 he spoke his *ignorabimus*. What this *ignorabimus* wants to say in essence is that on the one hand, we have in the historical evolution of humanity arrived at clarity regarding nature and have constructed the concept of matter. In this view of nature we have lost man—that is, ourselves. On the other hand we look down into consciousness. To this realm we want to apply that which has been most important in arriving at the contemporary explanation of nature. Consciousness rejects this lucidity. This mathematical clarity is entirely out of place. To be sure, we find man in a sense, but our consciousness is not yet strong enough, not yet intensive enough to comprehend man fully.

Again, one is tempted to answer with an *ignorabimus*, but that cannot be, for we need something more than an *ignorabimus* in order to meet the social demands of the modern world. The limit that Du Bois-Reymond had come up against when he spoke his *ignorabimus* on August 14, 1872 lies not within the human condition as such but only within its present stage of historical human evolution. How are we to transcend this *ignorabimus*? That is the burning question.

It must be answered, not to meet a human "need to know" but to meet man's universal need to become fully human. And in just what way one can strive for an answer, in what way the *ignorabimus* can be overcome to fulfill the demands of human evolution—this shall be the theme of our course of lectures as it proceeds.

II

To those who demand of a cycle of lectures with a title such as ours that nothing be introduced that might interfere with the objective presentation of ideas, I would like, since today I shall have to mention certain personalities, to say the following. The moment one begins to represent the results of human judgment in their relationship to life, to full human existence, it becomes inevitable that one indicate the personalities with whom the judgments originated. Even in a scientific presentation, one must remain within the sphere in which the judgment arises, within the realm of human struggling and striving toward such a judgment. And especially since the question we want above all to answer is: what can be gleaned from modern scientific theories that can become a vital social thinking able to transform thought into impulses for life?—then one must realize that the series of considerations one undertakes is no longer confined to the study and the lecture halls but stands rather within the living evolution of humanity.

Behind everything with which I commenced yesterday, the modern striving for a mathematical-mechanical world view and the dissolution of that world view, behind that which came to a climax in 1872 in the famous speech by the physiologist, Du Bois-Reymond, concerning the limits of natural science, there stands something even more important. It is something that begins to impress itself upon us the moment we want to begin to speak in a living way about the limits of natural science.

15

A personality of extraordinary philosophical stature still looks over to us with a certain vitality out of the first half of the nineteenth century: Hegel. Only in the last few years has Hegel begun to be mentioned in the lecture halls and in the philosophical literature with somewhat more respect than in the recent past. In the last third of the nineteenth century the academic world attacked Hegel outright, yet one could demonstrate irrefutably that Eduard von Hartmann had been quite right in claiming that during the 1880s only two university lecturers in all of Germany had actually read him. The academics opposed Hegel but not on philosophical grounds, for as a philosopher they hardly knew him. Yet they knew him in a different way, in a way in which we still know him today. Few know Hegel as he is contained, or perhaps better said, as his world view is contained, in the many volumes that sit in the libraries. Those who know Hegel in this original form so peculiar to him are few indeed. Yet in certain modified forms he has become in a sense the most popular philosopher the world has ever known. Anyone who participates in a workers' meeting today or, even better, anyone who had participated in one during the last few decades and had heard what was discussed there; anyone with any sense for the source of the mode of thinking that had entered into these workers' meetings, who really knew the development of modern thought, could see that this mode of thinking had originated with Hegel and flowed through certain channels out into the broadest masses. And whoever investigated the literature and philosophy of Eastern Europe in this regard would find that the Hegelian mode of thinking had permeated to the farthest reaches of Russian cultural life. One thus could say that, anonymously, as it were, Hegel has become within the last few decades perhaps one of the most influential philosophers in human history. On the other hand, however, when

16

one perceives what has come to be recognized by the broadest spectrum of humanity as Hegelianism, one is reminded of the portrait of a rather ugly man that a kind artist painted in such a way as to please the man's family. As one of the younger sons, who had previously paid little attention to the portrait, grew older and really observed it for the first time, he said: "But father, how you have changed!" And when one sees what has become of Hegel one might well say: "Dear philosopher, how you have changed!" To be sure, something extraordinary has happened regarding this Hegelian world view.

Hardly had Hegel himself departed when his school fell apart. And one could see how this Hegelian school appropriated precisely the form of one of our new parliaments. There was a left wing and a right wing, an extreme left and an extreme right, an ultra-radical wing and an ultra-conservative wing. There were men with radical scientific and social views, who felt themselves to be Hegel's true spiritual heirs, and on the other side there were devout, positive theologians who wanted just as much to base their extreme theological conservatism on Hegel. There was a center for Hegelian studies headed by the amiable philosopher, Karl Rosenkranz, and each of these personalities, every one of them, insisted that he was Hegel's true heir.

What is this remarkable phenomenon in the evolution of human knowledge? What happened was that a philosopher once sought to raise humanity into the highest realms of thought. Even if one is opposed to Hegel, it cannot be denied that he dared attempt to call forth the world within the soul in the purest thought-forms. Hegel raised humanity into ethereal heights of thinking, but strangely enough, humanity then fell right back down out of those heights. It drew on the one hand certain materialistic and on the other hand certain positive theological conclusions from Hegel's

17

thought. And even if one considers the Hegelian center headed by the amiable Rosenkranz, even there one cannot find Hegel's philosophy as Hegel himself had conceived it. In Hegel's philosophy one finds a grand attempt to pursue the scientific method right up into the highest heights. Afterward, however, when his followers sought to work through Hegel's thoughts themselves, they found that one could arrive thereby at the most contrary points of view.

Now, one can argue about world views in the study, one can argue within the academies, and one can even argue in the academic literature, so long as worthless gossip and barren cliques do not result. These offspring of Hegelian philosophy, however, cannot be carried out of the lecture halls and the study into life as social impulses. One can argue conceptually about contrary world views, but within life itself these contrary world views do not fight so peaceably. One must use just such a paradoxical expression in describing such a phenomenon. And thus there stands before us in the first half of the nineteenth century an alarming factor in the evolution of human cognition, something that has proved itself to be socially useless in the highest degree. With this in mind we must then raise the question: how can we find a mode of thinking that can be useful in social life? In two phenomena above all we notice the uselessness of Hegelianism for social life.

One of those who studied Hegel most intensively, who brought Hegel fully to life within himself, was Karl Marx. And what is it that we find in Marx? A remarkable Hegelianism indeed! Hegel up upon the highest peak of the conceptual world—Hegel upon the highest peak of Idealism —and the faithful student, Karl Marx, immediately transforming the whole into its direct opposite, using what he believed to be Hegel's method to carry Hegel's truths to their logical conclusions. And thereby arises historical materialism, which is to be for the masses the one world view that can enter into social life. We thus are confronted in the first half of the nineteenth century with the great Idealist, Hegel, who lived only in the spirit, only in his ideas, and in the second half of the nineteenth century with his student, Karl Marx, who contemplated and recognized the reality of matter alone, who saw in everything ideal only ideology. If one but takes up into one's feeling this turnabout of conceptions of world and life in the course of the nineteenth century, one feels with all one's soul the need to achieve an understanding of nature that will serve as a basis for judgments that are socially viable.

Now, if we turn on the other hand to consider something that is not so obviously descended from Hegel but can be traced back to Hegel nonetheless, we find still within the first half of the nineteenth century, but carrying over into the second half, the "philosopher of the ego," Max Stirner. While Karl Marx occupies one of the two poles of human experience mentioned yesterday, the pole of matter upon which he bases all his considerations, Stirner, the philosopher of the ego, proceeds from the opposite pole, that of consciousness. And just as the modern world view, gravitating toward the pole of matter, becomes unable to discover consciousness within that element (as we saw yesterday in the example of Du Bois-Reymond), a person who gravitates to the opposite pole of consciousness will not be able to find the material world. And so it is with Max Stirner. For Max Stirner, no material universe with natural laws actually exists. Stirner sees the world as populated solely by human egos, by human consciousnesses that want only to indulge themselves to the full. "I have built my thing on nothing" —that is one of Max Stirner's maxims. And on these grounds Stirner opposes even the notion of Providence. He says for example: certain moralists demand that we should

19

not perform any deed out of egoism, but rather that we should perform it because it is pleasing to God. In acting, we should look to God, to that which pleases Him, that which He commands. Why, thinks Max Stirner, should I, who have built only upon the foundation of ego-consciousness, have to admit that God is after all the greater egoist Who can demand of man and the world that all should be performed as it suits Him? I will not surrender my own egoism for the sake of a greater egoism. I will do what pleases *me*. What do I care for a God when I have *myself?*

One thus becomes entangled and confused within a consciousness out of which one can no longer find the way. Yesterday I remarked how on the one hand we can arrive at clear ideas by awakening in the experience of ideas when we descend into our consciousness. These dreamlike ideas manifest themselves like drives from which we cannot then escape. One would say that Karl Marx achieved clear ideas —if anything his ideas are too clear. That was the secret of his success. Despite their complexity, Marx's ideas are so clear that, if properly garnished, they remain comprehensible to the widest circles. Here clarity has been the means to popularity. And until it realizes that within such a clarity humanity is lost, humanity, as long as it seeks logical consequences, will not let go of these clear ideas.

If one is inclined by temperament to the other extreme, to the pole of consciousness, one passes over onto Stirner's side of the scale. Then one despises this clarity: one feels that, applied to social thinking, this clarity makes man into a cog in a social order modeled on mathematics or mechanics—but into that only, into a mere cog. And if one does not feel oneself cut out for just that, then the will that is active in the depths of human consciousness revolts. Then one comes radically to oppose all clarity. One mocks all clarity, as Stirner did. One says to oneself: what do I care about anything else? What do I care even about nature? I shall project my own ego out of myself and see what happens. We shall see that the appearance of such extremes in the nineteenth century is in the highest degree characteristic of the whole of recent human evolution, for these extremes are the distant thunder that preceded the storm of social chaos we are now experiencing. One must understand this connection if one wants at all to speak about cognition today.

Yesterday we arrived at an indication of what happens when we begin to correlate our consciousness to an external natural world of the senses. Our consciousness awakens to clear concepts but loses itself. It loses itself to the extent that one can only posit empty concepts such as "matter," concepts that then become enigmatic. Only by thus losing ourselves, however, can we achieve the clear conceptual thinking we need to become fully human. In a certain sense we must first lose ourselves in order to find ourselves again out of ourselves. Yet now the time has come when we should learn something from these phenomena. And what can one learn from these phenomena? One can learn that, although clarity of conceptual thinking and perspicuity of mental representation can be won by man in his interaction with the world of sense, this clarity of conceptual thinking becomes useless the moment we strive scientifically for something more than a mere empiricism. It becomes useless the moment we try to proceed toward the kind of phenomenalism that Goethe the scientist cultivated, the moment we want something more than natural science, namely Goetheanism.

What does this imply? In establishing a correlation between our inner life and the external physical world of the senses we can use the concepts we form in interaction with nature in such a way that we try not to remain within the natural phenomena but to think on beyond them. We are doing this if we do more than simply say: within the spectrum there appears the color yellow next to the color green, and on the other side the blues. We are doing this if we do not simply interrelate the phenomena with the help of our concepts but seek instead, as it were, to pierce the veil of the senses and construct something more behind it with the aid of our concepts. We are doing this if we say: out of the clear concepts I have achieved I shall construct atoms, molecules—all the movements of matter that are supposed to exist behind natural phenomena. Thereby something extraordinary happens. What happens is that when I as a human being confront the world of nature [see illustration], I use my concepts not only to create for myself a conceptual order within the realm of the senses but also to break through the boundary of sense and construct behind it atoms and the like. I cannot bring my lucid thinking to a halt within the realm of the senses. I take my lesson from inert matter, which continues to roll on even when the propulsive force has ceased. My knowledge reaches the world of sense, and I remain in-

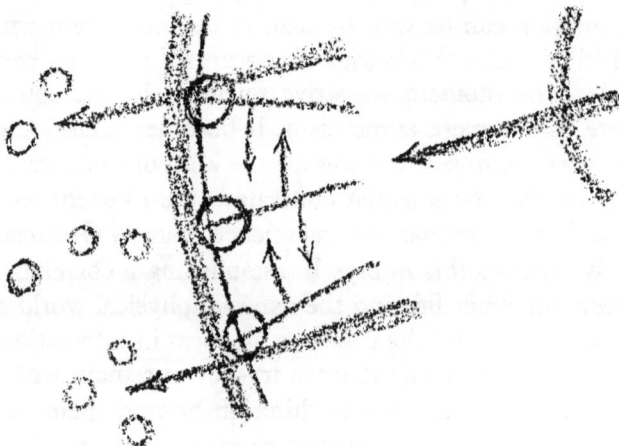

ert. I have a certain inertia, and I roll with my concepts on beyond the realm of the senses to construct there a world the existence of which I can begin to doubt when I notice that my thinking has only been borne along by inertia.

It is interesting to note that a great proportion of the philosophy that does not remain within phenomena is actually nothing other than just such an inert rolling-on beyond what really exists within the world. One simply cannot come to a halt. One wants to think ever farther and farther beyond and construct atoms and molecules—under certain circumstances other things as well that philosophers have assembled there. No wonder, then, that this web one has woven in a world created by the inertia of thinking must eventually unravel itself again.

Goethe rebelled against this law of inertia. He did not want to roll onward thus with his thinking but rather to come strictly to a halt at this limit [see illustration: heavy line] and to apply concepts within the realm of the senses. He thus would say to himself: within the spectrum appear to me yellow, blue, red, indigo, violet. If, however, I permeate these appearances of color with my world of concepts while remaining within the phenomena, then the phenomena order themselves of their own accord, and the phenomenon of the spectrum teaches me that when the darker colors or anything dark is placed behind the lighter colors or anything light, there appear the colors which lie toward the blue end of the spectrum. And conversely, if I place light behind dark, there appear the colors which lie toward the red end of the spectrum.

What was it that Goethe was actually seeking to do? Goethe wanted to find simple phenomena within the complex but above all such phenomena as allowed him to remain within this limit [see illustration], by means of which he did not roll on into a realm that one reaches only through a certain mental inertia. Goethe wanted to adhere to a strict phenomenalism. If we remain within phenomena and if we strive with our thinking to come to a halt there rather than allow ourselves to be carried onward by inertia, the old question arises in a new way. What meaning does the phenomenal world have when I consider it thus? What is the meaning of the mechanics and mathematics, of the number, weight, measure, or temporal relation that I import into this world? What is the meaning of this?

You know, perhaps, that the modern world conception has sought to characterize the phenomena of tone, color, warmth, etc. as only subjective, whereas it characterizes the so-called primary qualities, the qualities of weight, space, and time, as something not subjective but objective and inherent in things. This conception can be traced back principally to the English philosopher, John Locke, and it has to a considerable extent determined the philosophical basis of modern scientific thought. But the real question is: what place within our systematic science of nature as a whole do mathematics, do mechanics—these webs we weave within ourselves, or so it seems at first—what place do these occupy? We shall have to return to this question to consider the specific form it takes in Kantianism. Yet without going into the whole history of this development one can nonetheless emphasize our instinctive conviction that measuring or counting or weighing external objects is essentially different from ascribing to them any other qualities.

It certainly cannot be denied that light, tones, colors, and sensations of taste are related to us differently from that which we could represent as subject to mathematical-mechanical laws. For it really is a remarkable fact, a fact worthy of our consideration: you know that honey tastes sweet, but to a man with jaundice it tastes bitter—so we can say that we stand in a curious relationship to the qualities within this

24

realm—while on the other hand we could hardly maintain that any normal man would see a triangle as a triangle, but a man with jaundice would see it as a square! Certain differentiations thus do exist, and one must be cognizant of them; on the other hand, one must not draw absurd conclusions from them. And to this very day philosophical thinking has failed in the most extraordinary way to come to grips with this most fundamental epistemological question. We thus see how a contemporary philosopher, Koppelmann, over-trumps even Kant by saying, for example—you can read this on page 33 of his *Philosophical Inquiries [Weltanschauungsfragen]*: everything that relates to space and time we must first construct within by means of the understanding, whereas we are able to assimilate colors and tastes directly. We construct the icosahedron, the dodecahedron, etc.: we are able to construct the standard regular solids only because of the organization of our understanding. No wonder, then, claims Koppelmann, that we find in the world only those regular solids we can construct with our understanding. One thus can find Koppelmann saying almost literally that it is impossible for a geologist to come to a geometer with a crystal bounded by seven equilateral triangles precisely because—so Koppelmann claims—such a crystal would have a form that simply would not fit into our heads. That is out-Kanting Kant. And thus he would say that in the realm of the thing-in-itself crystals could exist that are bounded by seven regular triangles, but they cannot enter our head, and thus we pass them by; they do not exist for us.

Such thinkers forget but one thing: they forget—and it is just this that we want to indicate in the course of these lectures with all the forceful proofs we can muster—that the natural order governing the construction of our head also governs the construction of the regular polyhedrons, and it is for just this reason that our head constructs no other poly-

25

hedrons than those that actually confront us in the external world. For that, you see, is one of the basic differences between the so-called subjective qualities of tone, color, warmth, as well as the different qualities of touch, and that which confronts us in the mechanical-mathematical view of the world. That is the basic difference: tone and color leave us outside of ourselves; we must first take them in; we must first perceive them. As human beings we stand outside tone, color, warmth, etc. This is not entirely the case as regards warmth—I shall discuss that tomorrow—but to a certain extent this is true even of warmth. These qualities leave us initially outside ourselves, and we must perceive them. In formal, spatial, and temporal relationships and regarding weight this is not the case. We perceive objects in space but stand ourselves within the same space and the same lawfulness as the objects external to us. We stand within time just as do the external objects. Our physical existence begins and ends at a definite point in time. We stand within space and time in such a way that these things permeate us without our first perceiving them. The other things we must first perceive. Regarding weight, well, ladies and gentlemen, you will readily admit that this has little to do with perception, which is somewhat open to arbitrariness: otherwise many people who attain an undesired corpulence would be able to avoid this by perception alone, merely by having the faculty of perception. No, ladies and gentlemen, regarding weight we are bound up with the world entirely objectively, and the organization by means of which we stand within color, tone, warmth, etc. is powerless against that objectivity.

So now we must above all pose the question: how is it that we arrive at any mathematical-mechanical judgment? How do we arrive at a science of mathematics, at a science of mechanics? How is it, then, that this mathematics, this mechanics, is applicable to the external world of nature, and how is it

26

that there is a difference between the mathematical-mechanical qualities of external objects and those that confront us as the so-called subjective qualities of sensation, tone, color, warmth, etc.?

At the one extreme, then, we are confronted with this fundamental question. Tomorrow we shall discuss another such question. Then we shall have two starting-points from which we can proceed to investigate the nature of science. Thence we shall proceed to the other extreme to investigate the formation of social judgments.

III

We have seen that one arrives at two limits when one seeks either to penetrate more deeply into natural phenomena or, proceeding from the state of normal consciousness, to penetrate more deeply into one's own being in order to uncover the essential nature of consciousness. Yesterday we showed already what happens at the one limit to knowledge. We have seen that man awakes to full consciousness in coming into contact with an external, physical world of sense. Man would remain a more-or-less drowsy being, a being with a sleepy soul, if he could not awake in confronting external nature. And what has happened in the spiritual evolution of humanity, in man's gradual acquisition of knowledge about external nature, is actually nothing other than what happens every morning when we awake out of sleep or dream-consciousness by confronting an external world. This latter is a kind of moment of awakening, and in the course of the evolution of humanity we have to do with a gradual awakening, a kind of long, drawn-out moment of awakening.

Now, we have seen that at this frontier a certain inertia on the part of the soul very easily comes into play, so that when we come up against the extended world of phenomena we do not proceed in the manner of Goethean phenomenology by halting at this frontier and ordering the phenomena according to the representations, concepts, and ideas we have already gained, describing them in a systematic, rational manner, and so forth. Instead, we roll on a bit farther beyond the phenomena with our concepts and ideas and thereby create a world, for example a world of metaphysical atoms, molecules, and so forth. This world, when it is so constituted, is merely a fabrication of the mind, a world into which there enters a creeping doubt, so that we have to unravel again the theoretical web we have spun. And we have seen that it is possible to guard against such a violation of this frontier of our knowledge through phenomenalism, through working purely with the phenomena themselves. We have also had to show that at this point in our striving for knowledge something emerges that commends itself to our use as an immediate necessity: mathematics and that part of mechanics that can be comprehended without any empirical observation, i.e., the entire compass of so-called analytical mechanics.

If we call to mind everything comprehended by mathematics and analytical mechanics, we have before us the system of concepts that allows us to enter into phenomena with the utmost certainty. And yet, as I began to indicate yesterday, one should not deceive oneself, for the whole manner in which we call forth the notions of mathematics and analytical mechanics, this process within our souls, is entirely different from that employed when we experiment with or observe sensory data and then seek to comprehend them, when we try to gather knowledge from sensory experience. In order to arrive at the fullest clarity regarding these matters one must bring all one's mental energy to bear, for in this realm full clarity can be attained only with the greatest mental exertion.

What is the difference between accumulating knowledge from sensory experience in a Baconian manner and the more inward mode of apprehension we find in mathematics and analytical mechanics? One can sharply differentiate the latter from those modes of apprehension that are not inward in this way by formulating clearly the concepts of the parallelogram of motion and the parallelogram of forces. One theorem of analytical mechanics states that two angular vectors proceeding from one point result in a third vector. To say, however, that a vector of a specific force here [see diagram: a] and a vector of a specific force here [b] result in a third force, which can also be determined according to the parallelogram—that is another notion altogether.

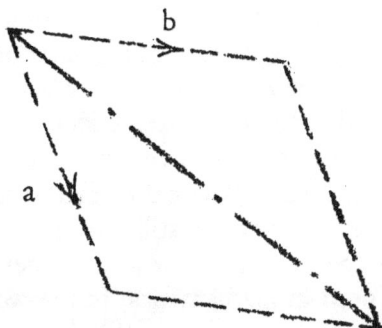

The parallelogram of motion lies strictly within the province of analytical mechanics, for it is internally consistent and demands no external proof. In this it is like the Rule of Pythagoras or any other geometrical axiom, but the existence of the parallelogram of forces can be determined only by experience, by experimentation. In this case, we bring something into that which we work through inwardly: the force that can be given only empirically from without. Here we no longer have a pure, analytical mechanics but an "empirical mechanics." One can thus differentiate sharply between that which is still actually mathematical—as we still conceive mathematics today—and that which leads over into conventional empiricism.

31

Now one stands before this phenomenon of mathematics as such. We comprehend mathematical truths. We proceed from mathematical phenomena to certain axioms. We weave the fabric of mathematics out of these axioms and then stand before an architectonic whole apprehended by the mind's eye [*im inneren Anschauen*]. If we are able by means of energetic thinking to differentiate sharply this inner apprehension from anything that can be experienced outwardly, we must see in this fabric of mathematics something that arises through an activity of soul entirely different from that which underlies our experience of the outer objects of sensation. Whether or not we arrive at a satisfactory comprehension of the world depends to a tremendous extent on our being able to make this clear distinction out of inner experience. We thus must ask: where does mathematics originate? Nowadays this question is still not pursued rigorously enough. One does not ask: how is this inner activity of the soul that we need in mathematics, in the wonderful architecture of mathematics—how is this inner activity of the soul different from that whereby we grasp external nature through the senses? One does not pose this question and seek an answer with sufficient rigor, because it is the tragedy of the materialistic world view that, while on the one hand it presses for sensory experience, on the other hand it is driven unawares into an abstract intellectualism, into a realm of abstraction where one is isolated from any true comprehension of the phenomena of the material world.

What kind of capacity is it, then, that we acquire when we engage in mathematics? We want to address ourselves to this question. In order to answer this question we must, I believe, have reached a complete understanding of one thing in particular: we must take fully seriously the concept of becoming as it applies to human life as well. We must begin by acquiring the discipline that modern science can

32

teach us. We must school ourselves in this way and then, taking the strict methodology, the scientific discipline we have learned from modern natural science, transcend it, so that we use the same exacting approach to rise into higher regions, thereby extending this methodology to the investigation of entirely different realms as well. For this reason I believe—and I want this to be expressly stated—that nobody can attain true knowledge of the spirit who has not acquired scientific discipline, who has not learned to investigate and think in the laboratories according to the modern scientific method. Those who pursue spiritual science [*Geisteswissenschaft*] have less cause to undervalue modern science than anyone. On the contrary, they know how to value it at its full worth. And many people—if I may here insert a personal remark—were extremely upset with me when, before publishing anything pertaining to spiritual science as such, I wrote a great deal about the problems of natural science in a way that appeared necessary to me. So you see it is necessary on the one hand for us to cultivate a scientific habit of mind, so that this can accompany us when we cross the frontiers of natural science. In addition, it is the quality of this scientific method and its results that we must take very seriously indeed.

You see, if we consider the simple phenomenon of warmth that appears when we rub two bodies together, it would be utterly unscientific to say, regarding this isolated phenomenon, that the warmth had been created *ex nihilo* or simply existed. Rather, we seek the conditions under which this warmth was previously latent and now appears by means of the bodies. We proceed from the one phenomenon to the other and thus take seriously this process of becoming [*das Werden*]. We must do the same with the concepts that we consider in spiritual science. So we must first of all ask: is that which manifests itself as the ability to perform math-

33

ematics present in man throughout his entire existence between birth and death? No, it is not always present. It awakes at a certain point in time. To be sure, we can, while still remaining empirical regarding the outer world, observe with great precision how there gradually arise out of the dark recesses of human consciousness faculties that manifest themselves as the ability to perform mathematics and something like mathematics that we have yet to discuss. If one can observe this emergence in time precisely and soberly, just as scientific research treats the phenomena of the melting or boiling point, one sees that this new faculty emerges at approximately that time of life when the child changes teeth. One must treat such a point in the development of human life with the same attitude with which physics, for example, teaches one to treat the melting or boiling point. One must acquire the ability to carry over into the complicated realm of human life the same strict inner discipline that one can acquire by observing simple physical phenomena according to the methods of modern science. If one does this, one sees that in the course of human development from birth, or rather from conception, up to the change of teeth, the soul faculties enabling one to perform mathematics manifest themselves gradually within the organism but that they are not yet fully present. Now we say that the warmth that manifests itself in a body under certain conditions was latent in that body beforehand, that it was at work within the inner structure of that body. In the same way we must be entirely clear that the capacity to perform mathematics, which becomes most evident at the change of teeth and reveals itself gradually in another sense, was also at work beforehand within the human organization. We thus arrive at an important and valuable insight into the nature of mathematics—mathematics taken, of course, in the very broadest sense. We begin to understand how that which is at our disposal after the change of teeth as a soul faculty worked previously within to organize us. Yes, within the child until approximately its seventh year there works an inner mathematics, an inner mathematics not abstract like our external one but full of active energy, a mathematics which, if I may use Plato's expression, not only can be inwardly envisioned [angeschaut] but is full of active life. Up to this point in time there exists within us something that "mathematicizes" us through and through.

When we ask at first entirely superficially what can be seen by looking empirically at this "latent mathematics" in the body of the young child, we are led to three things resembling inner senses. In the course of these lectures we shall come to see that one can indeed speak of senses within as well. Today I want only to indicate that we are led to something that develops an inward faculty of perception similar to the outward perception developed by the eyes and ears, except that the former remains unconscious within us during these first years. And if we look within, look into our own inner organization not like nebulous mystics but with all our powers of apprehension, we can find within three functions similar to those of the outward senses. We find inner senses that exercise a certain activity, a certain inner mathematics, just in those first several years. One encounters first of all what I would like to call the sense of life. This sense of life manifests itself in later years as a perception of our inner state as a whole. In a certain way we feel either well or unwell. We feel comfortable or uncomfortable: just as we have a faculty for perceiving outwardly with the eyes, so also do we have a faculty for perceiving inwardly. This faculty is directed toward the whole organism and is for that reason dark and dull; yet it is there all the same. We shall have more to say about this later. For the moment I want to anticipate this later discussion only by remarking that this

35

sense of life is—if I may use a tautology—especially active in the vitality of the child up until the change of teeth.

Another inner sense that we must consider when we look within in this way is that which I would like to call the sense of movement. We must form a clear conception of this sense of movement. When we move our limbs, we are aware of this not only by viewing ourselves externally but also by means of an internal perception. Also when we walk: we are conscious that we are walking not only in that we see objects pass and our view of the external world changes but also in that we have an internal perception of the movements of the limbs, of changes within ourselves as we move. Normally we remain unaware of the inner experiences and perceptions that run parallel to the outer because of the strength of the external impressions, much as a dim light is "extinguished" by a bright one.

And a third inward-looking faculty is the sense of balance. The sense of balance is what enables us to locate ourselves within the world, to avoid falling, to perceive in a certain way how we can bring ourselves into harmony with the forces in our environment. We perceive this process of bringing ourselves into harmony with our environment inwardly. We thus can truly say that we bear within ourselves these three inner senses: the sense of life, the sense of movement, and the sense of balance. They are especially active in childhood up to the change of teeth. Around this time of the change of teeth their activity begins to wane, but observe— to take but one example, the sense of balance—observe how at birth the child has as yet nothing enabling it to attain the position of balance it needs in later life. Consider how the child gradually gains control of itself, how it learns at first to crawl on all fours, how it gradually achieves through its sense of balance the ability to stand and to walk, how it finally is able to maintain its own balance.

36

If one considers the entire process of development from conception to the change of teeth, one sees therein the powerful activity of these three inner senses. And if one can attain a certain insight into what is happening there, one sees that there is at work in the sense of balance and the sense of movement nothing other than a living "mathematicizing" [*ein lebendiges Mathematisieren*]. In order for it to come to life, the sense of life is there to vitalize it. We thus see a kind of latent realm of mathematics active within man. This activity does not entirely cease at the change of teeth, but it does become at that time considerably less pronounced for the remainder of life. That which is inwardly active in the sense of balance, the sense of movement, and the sense of life becomes free. This latent mathematics becomes free, just as latent heat can become liberated heat. And we see how that which initially was woven through the organism as an element of soul becomes free. We see how this mathematics emerges as abstraction from a condition in which it was originally a concrete force shaping the human organism. And because as human beings we are suspended in the web of existence according to temporal and spatial relationships, we take this mathematics that has become free out into the world and seek to comprehend the external world by means of something that worked within us up until the change of teeth. You see, it is not a denial but rather an extension of natural science that results when one considers rightly what ought to live within spiritual science as attitude and will.

We thus carry what originates within ourselves beyond the frontier of sense perception. We observe man within a process of becoming. We do not simply observe mathematics on the one hand and sensory experience on the other but rather the emergence of mathematics within the developing human being. And now we come to that which truly leads over into spiritual science itself. You see, that which we call

37

forth out of our own inner life, this "mathematicizing," becomes in the end an abstraction. Yet our experience of it need not remain an abstraction. In our time there is, to be sure, little opportunity for us to experience mathematics in a true light. Yet at a certain point in the development of Western civilization there does come to light something of this sense of a special spirit in mathematics. This comes to light at the point where Novalis, the poet Novalis, who underwent a good mathematical training in his studies, writes about mathematics in his *Fragments*. He calls mathematics a grand poem, a wonderful, grand poem.

One really must have experienced at some time what it is that leads from an abstract understanding of the geometrical forms to a sense of wonder at the harmony that underlies this inner "mathematicizing." One really must have had the opportunity to get beyond the cold, sober performance of mathematics, which many people even hate. One must have struggled through as Novalis had in order to stand in awe of the inner harmony and—if I may use an expression you have heard often in a completely different context—the "melody" [*Melos*] of mathematics.

Then something new enters into one's experience of mathematics. There enters into mathematics, which otherwise remains purely intellectual and, metaphorically speaking, interests only the head, something that engages the entire man. This something manifests itself in such youthful spirits as Novalis in the feeling: that which you behold as mathematical harmony, that which you weave through all the phenomena of the universe, is actually the same loom that wove you during the first years of growth as a child here on earth. This is to feel concretely man's connection with the cosmos. And when one works one's way through to such an inner experience, which many hold to be mere fantasy because they have not actually attained it themselves, one has some idea what the spiritual scientist [*Geistesforscher*] experiences when he rises to a more extensive grasp of this "mathematicizing" by undergoing an inner development of which I have yet to speak and which you will find fully depicted in my book, *Knowledge of the Higher Worlds and its Attainment*.[2] For then the capacity of soul manifesting itself as this inner mathematics passes over into something far more comprehensive. It becomes something that remains just as exact as mathematical thought yet does not proceed solely from the intellect but from the whole man.

On this path of constant inner work—an inner work far more demanding than that performed in the laboratory or observatory or any other scientific institution—one comes to know what it is that underlies mathematics, that underlies this simple faculty of the human soul which can be expanded into something far more comprehensive. In this higher experience of mathematics one comes to know Inspiration. One comes to understand the differences between what lives in us as mathematics and what lives in us as outer-directed empiricism. In this outer-directed empiricism we have sense impressions that give content to our empty concepts. In Inspiration we have something inwardly spiritual, the activity of which manifests itself already in mathematics, if we know how to grasp mathematics properly—something spiritual which in our early years lives and weaves within us. This activity continues. In doing mathematics we experience this in part. We come to realize that the faculty for performing mathematics rests upon Inspiration, and we can come to experience Inspiration itself by evolving into spiritual scientists. Our representations and concepts then receive their content in a way other than through external experience. We can inspire ourselves with the spiritual force that works within us during childhood. For what works within us during our childhood is spirit. The spirit, however, resides in the human body and must be perceived there through the body, within man. It can be viewed in its pure, free form if one acquires through the faculty of Inspiration the capacity not only to think in mathematical concepts but to view that which exists as a real force in that it organizes us through and through up until the seventh year. And that which manifests itself partially in mathematics and reveals itself as a much more expansive realm through Inspiration can be inwardly viewed, if one employs certain spiritual scientific methods about which—as I have said—I plan yet to speak. One thereby gains not merely new results to add to those acquired through the old powers of cognition but rather an entirely new mode of apprehension. One acquires a new "Inspirative" cognition.

The course of human evolution has been such that these powers of Inspirative cognition have receded with the passage of time, after having been present earlier to a very high degree. One must come to understand how Inspiration arises within the inner being of man—that same Inspiration that survives in the West only in the diluted, intellectual experience of mathematics. The experience can be expanded, however, if only one comprehends fully the inner nature of that realm; only then does one begin to understand what lived in that earlier consciousness transmitted to us actually only from the East, from the Vedanta and the other Eastern philosophies that remain so cryptic to the Western mind. For what was it that actually lived within these Eastern philosophies? It was something that arose through soul faculties of a mathematical nature. It was an Inspiration. It was not merely mathematics but rather something attained within the soul in a way similar to that in which one performs mathematics. Thus I would say that the mathematical atmosphere emanating from the Vedanta and similar ancient world views is something that can be understood from the perspective one attains in rising again to enter the realm of Inspiration. If one can raise to vivid inner life that which works unconsciously in mathematics and the mathematical sciences and can carry it over into another realm, one discovers the same mathematical element that Goethe viewed. Goethe modestly confessed that he did not have proficiency in mathematics in any conventional sense. Goethe has written on his relationship to mathematics in a very interesting series of essays, which you can find in his scientific writings under the heading "Relationship to Mathematics." Extraordinarily interesting! For despite Goethe's modest confession that he had not acquired a proficiency in the handling of actual mathematical concepts and theories, he does require one thing: he calls for a phenomenalism such as he employed in his own scientific studies. He demands that within the secondary phenomena confronting us in the phenomenal world we seek the archetypal phenomenon [*Urphänomen*]. But just what kind of activity is this? He demands that we trace external phenomena back to the archetypal phenomenon, in just the same way that the mathematician traces the outward apprehension [*äusseres Anschauen*] of complex structures back to the axiom. Goethe's archetypal phenomena are empirical axioms, axioms that can be experienced.

Goethe thus demands, in a truly mathematical spirit, that one inwardly permeate phenomena with mathematics. He writes that we must see the archetypal phenomena in such a way that we are able at all times to justify our procedures according to the rigorous requirements of the mathematician. Thus what Goethe seeks is a modified, transformed mathematics, one that suffuses phenomena. He demands this as a scientific activity.

Goethe was able, therefore, to suffuse with light the one pole that otherwise remains so dark if we postulate only the concept of matter. We shall see how Goethe approached this

41

pole; we moderns must, however, approach the other, the pole of consciousness. We must investigate in the same way how soul faculties manifest their activity in the human being, how they proceed from man's inner nature to manifest their activity externally. We shall have to investigate this. It shall become clear that we must complement the method of investigating the external world offered by Goethean phenomenology with a method of comprehending the realm of human consciousness. It must be a mode of comprehension justifiable in the sense in which Goethe's can be justified to the mathematician—a method such as I tried to employ in a modest way in my book, *The Philosophy of Freedom*.[3]

At the pole of matter we thus encounter the results yielded by Goethean phenomenology and at the pole of consciousness those attained by pursuing the method that I sought to establish in a modest way in my *Philosophy of Freedom*.

Tomorrow we will want to pursue this further.

IV

Yesterday's considerations led us to conclude that at one boundary of cognition we must come to a halt within phenomena and then permeate them with what the phenomena call forth within our consciousness, with concepts, ideas, and so forth. It became apparent that the realm in which these ideas are most pure and pellucid is that of mathematics and analytical mechanics. Our considerations then climaxed in showing how reflection reveals that everything present in the soul as mathematics, as analytical mechanics, actually rests upon Inspiration. Then we were able to indicate how the impulses proceeding from Inspiration are diffused throughout the ancient Indian Vedanta: the same spirit from which we now draw only mathematics and analytical mechanics was once the source of the delicate spirituality of the Vedanta. We were able to show how Goethe, in establishing his mode of phenomenology, always strives to find the archetypal phenomenon while remaining within the phenomena themselves and that his search for the archetypal phenomenon that underlies complex phenomena is, inwardly, the same as the mathematician's search for the axiom underlying complex mathematical constructs. Goethe, therefore, who himself admitted that he had no conventional mathematical training, nevertheless sensed the essence of mathematics so clearly that he demanded a method for the determination of archetypal phenomena rigorous enough to satisfy a mathematician. It is just this that the Western mind

43

finds so attractive in the Vedanta: that in its inner organization, in its progression from one contemplation to the next, it reveals the same inner necessity as mathematics and analytical mechanics. That such connections are not uncovered by academic studies of the Vedanta is simply a consequence of there being so few people today with a universal education. Those who engage in pursuits that then lead them into Oriental philosophy have too little comprehension—and, as I have said, Goethe did have this—of the true inner structure of mathematics. They thus never grasp this philosophy's vital nerve. At the one pole, then, the pole of matter, we have been able to indicate the attitude we must assume initially if we do not wish to continue weaving a Penelope's web like the world view woven by recent science but rather to come to grips with something that rests upon a firm foundation, that bears its center of gravity within itself.

On the other side there stands, as I indicated yesterday, the pole of consciousness. If we attempt to investigate the content of consciousness merely by brooding our way into our souls in the nebulous manner of certain mystics, what we attain are actually nothing but certain reminiscences that have been stored up in our consciousness since birth, since our childhoods. This can easily be demonstrated empirically. One need think only of a certain man well educated in the natural sciences who, in order to demonstrate that the so-called "inner life" partakes of the nature of reminiscences, describes an experience he once had while standing in front of a bookstore. In the store he saw a book that captured his attention by its title. It dealt with the lower forms of animal life. And, seeing this book, he had to smile. Now imagine how astonished he was: a serious scientist, a professor, who sees a book title in a bookstore—a book on the lower animals at that!—and feels compelled to smile! Then he began to ponder just whence this smile might have come.

At first he could think of nothing. And then it occurred to him: I shall close my eyes. And as he closed his eyes and it became dark all around him, he heard in the distance a musical motif. Hearing this musical motif in the moment reminded him of the music he had heard as a young lad when he danced for the first time. And he realized that of course there lived in his subconscious not only this musical motif but also a bit of the partner with whom he had hopped about. He realized how something that his normal consciousness had long since forgotten, something that had not made so strong an impression on him that he would have thought it possible for it to remain distinct for a whole lifetime, had now risen up within him as a whole complex of associations. And in the moment in which his attention had been occupied with a serious book, he had not been conscious that in the distance a music box was playing. Even the sounds of the music box had remained unconscious at the time. Only when he closed his eyes did they emerge.

Many things that are mere reminiscences emerge from consciousness in this way, and then some nebulous mystics come forth to tell us how they have become aware of a profound connection with the divine "Principle of Being" within their own inner life, how there resounds from within a higher experience, a rebirth of the human soul. And thereby vast mystical webs are woven, webs that are nothing but the forgotten melody of the music box. One can ascribe a great deal of the mystical literature to this forgotten melody of the music box.

This is precisely what a true spiritual science requires: that we remain circumspect and precise enough to refrain from trumpeting forth everything that arises out of the unconscious as reminiscences, as mysticism, as though it were something that could lay claim to objective meaning. For it is just the spiritual scientist who most needs inner clarity if he wishes to work in a truly fruitful way in this direction. He needs inner clarity above all when he undertakes to delve into the depths of consciousness in order to come to grips with its true nature. One must delve into the depths of consciousness itself, yet at the same time one must not remain a dilettante. One must acquire a professional competence in everything that psychopathology, psychology, and physiology have determined in order to be able to differentiate between that which makes an unjustifiable claim to spiritual scientific recognition and that which has been gained through the same kind of discipline, as, for example, mathematics or analytical mechanics.

To this end I sought already in the last century to characterize in a modest way this other pole, the pole of consciousness, as opposed to the pole of matter. To understand the pole of matter requires that we build upon Goethe's view of nature. The pole of consciousness, on the other hand, was not to be reached so easily by a Goetheanistic approach, for the simple reason that Goethe was no trivial thinker, nor trivial in his feelings when it was a matter of cognition. Rather, he brought with him into this realm all the reverence that is necessary if one seeks to approach the springs of knowledge. And thus Goethe, who was by disposition more attuned to external nature, felt a certain anxiety about anything that would lead down into the depths of consciousness itself, about thinking elaborated into its highest, purest forms. Goethe felt blessed that he had never thought about thinking. One must understand what Goethe meant by this, for one cannot actually think about thinking. One cannot actually think thinking any more than one can "iron" iron or "wood" wood. But one can do something else. What one can do is attempt to follow the paths that are opened up in thinking when it becomes more and more rational, to pursue them in the way one does through the discipline of

46

mathematical thinking. If one does this, one enters via a natural inner progression into the realm that I sought to consider in my *Philosophy of Freedom*. What one attains in this way is not a thinking about thinking. One can speak of thinking about thinking in a metaphorical sense at best. One does attain something else, however: what one attains is an actual viewing [*Anschauen*] of thinking, but to arrive at this "viewing of thinking," it is necessary first to have acquired a concrete notion of the nature of sense-free thinking. One must have progressed so far in the inner work of thinking that one attains a state of consciousness in which one recognizes one's thinking to be sense-free merely by grasping that thinking, by "viewing" it as such.

This is the path that I sought to follow—if only, as I have said, in a modest way—in my *Philosophy of Freedom*. What I sought there was first to make thinking sense-free and then to present this thinking to consciousness in the same way that mathematics or the faculties and powers of analytical mechanics are present to consciousness when one pursues these sciences with the requisite discipline.

Perhaps at this juncture I might be allowed to add a personal remark. In positing this sense-free thinking as a simple fact, yet nevertheless a fact capable of rigorous demonstration in that it can be called forth in inner experience like the structure of mathematics, I flew in the face of every kind of philosophy current in the 1880s and 1890s. It was objected again and again: this "sense-free thinking" has no basis in any kind of reality. Already in my *Theory of Knowledge Based on Goethe's World Conception*,[4] however, in the early 1880s, I had pointed to the experience of pure thinking, in the presence of which one realizes: you are now living in an element that no longer contains any sense impressions and nevertheless reveals itself in its inner activity as a reality. Of this thinking I had to say that it is here we find the true spiritual communion of humanity and union with reality. It is as though we have grabbed the coattails of universal being and can feel how we are related to it as souls. I had to protest vigorously against what was then the trend in philosophy, that to which Eduard von Hartmann paid homage in 1869 by giving his *Philosophy of the Unconscious* the motto: "Speculative Results Following the Method of Scientific Induction." That was a philosophical bow to natural science. I wrote to protest against this insubstantial metaphysics, which arises only when we allow our thinking to roll on beyond the veil of sense as I have described. I thus gave my *Philosophy of Freedom* the motto: "Observations of the Soul According to the Scientific Method." I wished to indicate thereby that the content of a philosophy is not contrived but rather in the strictest sense the result of inner observation, just as color and sound result from observation of the outer world. And in experiencing this element of pure thought—an element that, to be sure, has a certain chilling effect on human nature—one makes a discovery. One discovers that human beings certainly can speak instinctively of freedom, that within man there do exist impulses that definitely tend toward freedom but that these impulses remain unconscious and instinctive until one rediscovers freedom in one's own thinking. For out of sense-free thinking there can flow impulses to moral action which, because we have attained a mode of thinking that is devoid of sensation, are no longer determined by the senses but by pure spirit. One experiences pure spirit by observing, by actually observing how moral forces flow into sense-free thinking. What one gains in this way above all is that one is able to bid farewell to any sort of mystical superstition, for superstition results in something that is in a way hidden and is only assumed on the basis of dark intimations. One can bid it farewell because now one has experienced in one's con-

48

sciousness something that is inwardly transparent, something that no longer receives its impulses from without but fills itself from within with spiritual content. One has grasped universal being at one point in making oneself exclusively a theater of cognition; one has grasped the activity of universal being in its true form and observed how it yields itself to us when we give ourselves over to this inner contemplation. We grasp the actuality of universal being at one point only. We grasp it not as abstract thought but as a reality when moral impulses weave themselves into the fabric of sense-free thinking. These impulses show themselves to be free in that they no longer live as instinct but in the garb of sense-free thinking. We experience freedom—to be sure a freedom that we realize immediately man can only approach in the way that a hyperbola approaches its asymptote, yet we know that this freedom lives within man to the extent that the spirit lives within him. We first conceive the spirit within the element of freedom.

We thereby discover something deep within man that weaves together the impulses of our moral-social actions— freedom—and cognition, that which we finally attain scientifically. By grasping freedom within sense-free thinking, by understanding that this comprehension occurs only within the realm of spirit, we experience that while performing this we are indeed within the spirit. We experience a mode of cognition that manifests itself simultaneously as an inner activity. It is an inner activity that can become a deed in the external world, something entirely capable of flowing over into the social life. At that time I sought to make two points absolutely clear, but at that time they were hardly understood. I tried above all to make clear that the most important thing about following such a cognitional path is the inner "schooling" [*Erziehung*] that we undertake. Yes, to have attained sense-free thinking is no small thing. One

49

must undergo many inner trials. One must overcome obstacles of which otherwise one has hardly any idea. By overcoming these obstacles; by finally attaining an inner experience that can hardly be retained because it escapes normal human powers so easily; by immersing oneself in this essence, one does not proceed in a nebulous, mystical way, but rather one descends into a luminous clarity, one immerses oneself in spirit. One comes to know the spirit. One knows what spirit is, knows because one has found the spirit by traveling along a path followed by the rest of humanity as well, except that they do not follow it to its end. It is a path, though, that must be followed to its end by all those who would strive to fulfill the social and cognitional needs of our age and to become active in those realms. That is the one thing that I intimated in my *Philosophy of Freedom.*

The other thing I intimated is that when we have found the freedom that lives in sense-free thinking to be the basis of true morality, we can no longer seek to deduce moral concepts and moral imperatives as a kind of analogue of natural phenomena. We must renounce everything that would lead us to ethical content obtained according to the method of natural science; this ethical content must come forth freely out of supersensible experience. I ventured to use a term that was little understood at the time but that absolutely must be posited if one enters this inner realm and wishes to understand freedom at all. I expressed it thus: the moral realm arises within us in our moral imagination [*moralische Phantasie*]. I employed this term "moral imagination" with conscious intent in order to indicate that—just as with the creations of the imagination [*Phantasie*]—the requisite spiritual effort is expended within man, regardless of anything external, and to indicate on the other hand that everything that makes the world morally and religiously valuable for us —namely moral imperatives—can be grasped only within

50

this realm that remains free from all external impressions and has as its ground man's inner activity alone. At the same time I indicated clearly in my *Philosophy of Freedom* that, if we remain within human experience, moral content reveals itself to us as the content of moral imagination but that when we enter more deeply into this moral content, which we bear down out of the spiritual world, we simultaneously enter the external world of the senses.

If you really study this philosophy, you shall see clearly the door through which it offers access to the spirit. Yet in formulating it I proceed in such a way that my method could meet the rigorous requirements of analytical mechanics, and I placed no value on any concurrence with the twaddle arising out of spiritualism and nebulous mysticism. One can easily earn approbation from these sides if one wants to ramble on idly about "the spirit" but avoids the inner path that I sought to traverse at that time. I sought to bring certainty and rigor into the investigation of the spirit, and it remained a matter of total indifference to me whether my results concurred with all the twaddle that comes forth from nebulous mystical depths to represent the spirit. At the same time, however, something else was gained in this process.

If one pursues further the two paths that I described on the basis of actual observation of consciousness in my *Philosophy of Freedom*, if one goes yet further, takes the next step—then what? If one has attained the inner experiences that are to be found within the sphere of pure thought, experiences that reveal themselves in the end as experiences of freedom, one achieves a transformation of the cognitional process with respect to the inner realm of consciousness. Then concepts and ideas no longer remain merely that; Hegelianism no longer remains Hegelianism and abstraction no longer abstraction, for at this point consciousness passes over into the actual realm of the spirit. Then one's immedi-

51

ate experience is no longer the mere "concept," the mere "idea," no longer the realm of thought that constitutes Hegelian philosophy—no: now concepts and ideas transform themselves into images, into Imagination. One discovers the higher plane of which moral imagination is only the initial projection; one discovers the cognitional level of Imagination. While philosophizing, one remains caught within a self-created reality; now, after pursuing the inner path indicated by my *Philosophy of Freedom*, after transcending the level of imagination [*Phantasie*], one enters a realm of ideas that are no longer dream-images but are grounded in spiritual realities, just as color and tone are grounded in the realities of sense. At this point one attains the realm of Imagination, a thinking in pictures [*bildliches Denken*]. One attains Imaginations that are real, that are no longer merely a subjective inner experience but part of an objective spiritual world. One attains Inspiration, which can be experienced when one performs mathematics in the right way, when this performance of mathematics itself becomes an experience that can then be developed further into that which one finds in the Vedanta. Inspiration is complemented at the other pole by Imagination, and only through Imagination does one arrive at something enabling one to comprehend man. In Imaginations, in pictorial representations [*bildhafte Vorstellungen*]—representations that have a more concrete content than abstract thoughts—one finds what is needed to comprehend man from the point of view of consciousness. One must renounce proceeding further when one has reached this point and not simply allow sense-free thinking to roll on with a kind of inner inertia, nor believe that one can penetrate into the secret depths of consciousness through sense-free thinking. Instead one must have the resolve to call a halt and confront the "external world" of the spirit from within. Then one will no longer

52

spin thoughts into a consciousness that can never fully grasp them; rather, one will receive Imagination, through which consciousness can finally be comprehended. One must learn to call a halt at this limit within the phenomena themselves, and thoughts then reveal themselves to one as that within cognition which can organize these phenomena; one needs to renounce at the outward limit of cognition and thereby receive the spiritual complement to phenomena in the intellect. In just this way one must renounce in the process of inner investigation, one must come to a halt with one's thinking and transform it. Thinking must be brought inwardly to a kind of reflection [*Reflexion*] capable of receiving images that then unfold the inner nature of man. Let me indicate the soul's inner life in this way [see illustration]. If through self-contemplation and sense-free thinking I approach this inner realm, I must not roll onward with my thinking lest I pass into a region where sense-free thinking finds nothing and can call forth only subjective pictures or

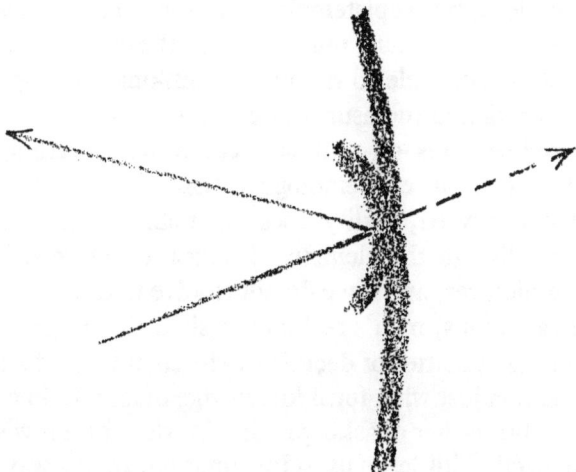

reminiscences out of my past. I must renounce and turn back. But then Imagination will reveal itself at the point of reflection. Then the inner world reveals itself to me as a world of Imagination.

Now, you see, we arrive inwardly at two poles. By proceeding into the outer world we approach the pole of Inspiration; by proceeding into the inner world of consciousness we approach the pole of Imagination. Once one has grasped these Imaginations it becomes possible to collate them, just as one collates data concerning external nature by means of experiments and conceptual thinking. In this manner one can collate inwardly something real, something that is not a physical body but an etheric body informing man's physical body throughout his whole life, yet in an especially intensive manner during the first seven years. At the change of teeth this etheric body takes on a somewhat different configuration [*Gestalt*], as I described to you yesterday. By having attained Imagination one is able to observe the way in which the etheric or life-body works within the physical body.

Now, it would be easy to object from the standpoint of some philosophical epistemology or other: if he wishes to remain logical, man must remain within the conceptual, within what is accessible to discursive thinking and capable of demonstration in the usual sense of the term. Fine. One can philosophize thus on and on. Yet however strong one's belief in such an epistemological tissue, however logically correct it may be, reality does not manifest itself thus; it does not live in the element of logical constructs. Reality lives in pictures, and if we do not resolve to achieve pictures or Imaginations, man's real nature shall elude our grasp. It is not at all a matter of deciding beforehand out of a certain predilection just what form knowledge must take in order to be valid but rather of asking reality in what form it wishes to reveal itself. This leads us to Imagination. In this way, then,

54

what lives within moral imagination manifests itself as the projection into normal consciousness of a higher spiritual world that can be grasped in Imagination.

And thus, ladies and gentlemen, I have led you, or at least sought to lead you, to the two poles of Inspiration and Imagination, which we shall consider more closely in the next few days in the light of spiritual science. I had to lead you to the portal, as it were, beforehand, in order to show that the existence of this portal is well founded in the normal scientific sense. For it is only upon such a foundation that we later can build the edifice of spiritual science itself, which we enter through that portal. To be sure, in traversing the long path, in employing the extremely demanding epistemological method I described to you today—which many may feel is difficult to understand—one must have the courage to come to grips not only with Hegel but also with "anti-Hegel." One must not only pursue the Hegelianism that I sought to depict in my *Riddles of Philosophy*[5]; one must also learn to give Stirner his due, for in Stirner's philosophy there lies something that rises out of consciousness to reveal itself as the ego. And if one simply gives rein to this ego that comes forth out of instinctive experiences, if one does not permeate it with that which manifests itself as moral imagination and Imagination, this ego becomes antisocial. As we have seen, *The Philosophy of Freedom* attempts to replace Stirner's egoism with something truly social. One must have the courage to pass through the instinctive ego Stirner describes in order to reach Imagination, and one must also have the courage to confront face-to-face the psychology of association that Mill, Spencer, and other likeminded proponents have sought to promulgate, a psychology that seeks to comprehend consciousness in a bare concept but cannot. One must have the courage to realize and admit to oneself that today we must follow another path entirely. The ancient Oriental could follow a path no longer accessible to us, in that he formulated his experiences of an inner mathematics in the Vedanta. This path is no longer accessible to the West. Humanity is in a process of constant evolution. It has progressed. Another path, another method, must be sought. This new method is now in its infancy, and its immaturity is best revealed when one realizes that this psychology of association, which seeks to collate inner representations according to laws in the same way one collates the data of natural phenomena, is nothing but the inertia of thinking that wants to break through a boundary but actually enters a void. To understand this one must come to know this psychology of association for what it really is and then learn to lead it over through an inner contemplative viewing [*Schauung*] into the realm of Imagination. Just as the Orient once saw the Vedanta arise within an element of primal mathematical thought and was able to enter thus into the spirituality of the external world, so we must seek the spirit in the way in which it tasks us today: we must look within and have the courage to proceed from mere concepts and ideas to Imaginations, to develop this pictorial consciousness within and thereby to discover the spirituality within ourselves. Then we shall be able to bear this spirituality back out into the external world. We shall have attained a spirituality grasped by the inner being of man, a spirituality that thus can bear fruit within the social life. The quality of our social life shall depend entirely on our nurturing a mode of cognition such as this, which can at the same time embrace the social. That this is the case I hope to show in the lectures yet to follow.

V

Today it will be necessary to come to terms with a number of things that actually can be understood only if one is able to overcome certain prejudices that have long been cultivated and zealously inculcated right up to the present day. Much of what shall be said here today, and further substantiated tomorrow, must be comprehended through raising oneself up to an inner viewing [*Anschauung*] of the spirit. You must consider that when the results of a scientific investigation of the spirit are met with a demand for proof such as is recognized by contemporary science or jurisprudence, or even contemporary social science—which is so useless in the face of life itself—one does not get very far at all. For the true spiritual scientist must already bear this method of demonstration within himself. He must have schooled himself in the rigorous methods of contemporary science, even of the mathematical sciences. He must know what mode of demonstration is demanded in these circles, and he must suffuse the processes of his whole inner life with this method: therein he builds the foundation for a higher mode of cognition. For this reason it is usually the case that when the demands of normal consciousness are placed before the spiritual scientist, he is thoroughly at home in the field from which the question stems. He has long since anticipated the objections that can be raised. One could even go so far as to say that he is only a spiritual scientist in the true sense of the word—in the sense in which we characterized spiritual science yesterday—to the extent that he has subjected himself to the rigorous discipline of the modern scientific method and knows at least the tenor of modern scientific thought quite well. I must make this one preliminary remark and add one other. If one cannot transcend the manner of demonstration that experimentation has made scientific habit, one shall never attain knowledge that can benefit society. For in a scientific experiment one proceeds—even if one cherishes the illusion that it is otherwise—in such a way that one moves in a certain direction and allows phenomena to confirm what lives within the ideas one has formulated as a natural law, or perhaps mathematically.

Now, when one is required to translate one's knowledge into social judgments, in other words, if the ideas that one has formulated as the natural laws of contemporary anthropology or biology or Darwinism—no matter how "progressive" this Darwinism might be—are to have validity; if one wants to translate them into a social science that can become truly practical, this knowledge obtained through experimentation is totally inadequate. It is totally inadequate because one cannot simply sit in a laboratory and wait to see what one's ideas call forth when they are applied to society. Thereby thousands upon thousands of people could easily die or starve or be made to suffer in some other way. A great part of the misery in our society has been called forth in just this way. Because they have originated in pure experimentation, our ideas have gradually become too narrow and impoverished to subsist in reality, which they must be able to do if thought is ever to enrich the sphere of practical life. I have already indicated the stance the spiritual scientist must take regarding the two boundaries that arise within cognition—the boundaries at the poles of matter and consciousness—if he is to attain knowledge that can reflect light back into nature and at the same time forward into the social

58

future. I have shown that at the boundary of the material world one must not allow one's thinking to roll on with its own inertia in order to construct mechanistic, atomistic, or molecular world conceptions tending toward the metaphysical but call a halt at the boundary and develop instead something that normally is not yet present as a faculty of cognition. One must develop Inspiration. On the other hand, I have shown you that if one wishes to come to an understanding of consciousness, one must not attempt, as Anglo-American associative psychology does, to penetrate into consciousness with ideas and concepts called forth by the natural world. It must be entirely clear in one's mind that consciousness is constituted such that these ideas culled from the external world can gain no access. We must abandon such ideas and seek rather to enter the realm of Imaginative cognition. In order to achieve self-knowledge we must permeate the concepts and ideas with content, so that they become images. Until the view of man which was born in the West and now has all of civilization in its grasp is transformed into Imaginative cognition, we shall never progress in coming to terms with this second boundary presenting itself to normal human cognition.

At the same time, however, one can say that humanity has evolved from certain stages, now become historical, to the point that requires that it progress to Inspiration on the one hand and Imagination on the other. Whoever is able to perceive what humanity is undergoing at the present, what is just beginning to reveal its first symptoms, knows that forces are rising out of the depths of human evolution that tend toward the proper introduction of Imagination and Inspiration into human evolution.

Inspiration cannot be attained except by exercising a certain faculty of mental representation in the way that I described in my book, *Knowledge of the Higher Worlds and its Attainment*, and shall describe at least in outline in the coming lectures. When one has progressed far enough in a kind of inner self-cultivation, a schooling of the self in a certain form of mental representation [*Vorstellen*]; when one schools onself to live within the realm of representations, ideas, and concepts that live within the mind—then one learns what it means to live in Inspiration. For when one exercises consciously the faculty that otherwise "mathematicizes" within us during the first seven years up to the change of teeth (in normal life and in conventional science this occurs unconsciously), when one enters into this "living mathematics," into this "living mechanics," it is as though one were to fall asleep, entering not into unconsciousness or nebulous dreams but into a new form of consciousness that I shall begin to describe to you today. One takes up into full consciousness what otherwise works within as the sense of balance, the sense of movement, and the sense of life. It is as though one were to wrest from oneself what otherwise lives within as sensations of balance, movement, and life so that one lives within them with the extended mathematical representations. Tomorrow I shall speak about this at greater length. One passes over into another consciousness, within which one experiences something like a toneless weaving in a cosmic music. I cannot describe it otherwise. One unites with this weaving in a toneless music in a way similar to that by which one makes the physical body one's own through the activity of the ego in childhood. This weaving in a toneless music provides the other, rigorously demonstrable awareness that one is now outside the body with one's soul-spirit. One begins to comprehend that even in normal sleep one's soul-spirit is outside the body. Yet the experience of sleep is not permeated with that which vibrates when leaving the body consciously through one's own initiative, and one experiences initially something like an inner unrest, an inner unrest that exhibits a musical quality when one enters into it with full consciousness. This unrest is gradually elucidated when the musical element one experiences there becomes a kind of wordless revelation of speech from the spiritual cosmos. These matters naturally appear grotesque and paradoxical to those who hear them for the first time. Yet much has arisen in the course of cosmic evolution that first appeared paradoxical and grotesque, and human evolution will not advance if one wishes to pass over these phenomena only half-consciously or unconsciously. Initially one has only a certain experience, an experience of a kind of toneless music. Then out of this experience of toneless music there arises something which, when experienced, enables us to comprehend inwardly a content as meaningful as that which is conveyed to us when we listen outwardly to a man who speaks to us via sensible words. The spiritual world simply begins to speak, and one has only to begin to acquire an experience of this.

Then one comes to experience something at a higher level. One no longer only weaves and lives in a toneless music and no longer merely perceives the speech of the supersensible spiritual world: one begins to recognize the contours of something that reveals itself within this supersensible world, the contours of beings. Within this universal spiritual speech that one initially encounters there emerge individual spiritual beings, in the same way that we, listening at a lower level to the speech of another man, crystallize or organize—if I may use such trivial expressions—what reveals itself as his soul and spirit into something substantial. We begin to live within the contemplation and knowledge of a spiritual reality. This realm of the spirit replaces the vacuous, insubstantial, metaphysical world of atoms and molecules: it confronts us as the reality that lies behind the phenomena of the sense world. We no longer stand in the same relation to the boundary of the material world as when we allow conceptualizing to roll on with its own inertia, attempting to carry the kind of thinking developed through interaction with the sense world beyond the boundary. Now we stand in a relationship to this boundary of sense such that the spiritual content of the world suddenly stands revealed there. This is one boundary to cognition.

Ladies and gentlemen, humanity at this point in its evolution is yearning to step out of itself, to step out of the body in this way, and one can see this tendency exemplified quite clearly in certain individuals. Human beings seek to withdraw from their bodies that which the spiritual scientist withdraws with full consciousness. The spiritual scientist withdraws this in a way analogous to the way in which he applies inwardly obtained concepts in a systematic, organized fashion to the natural world. As some of you will know, for some time now a great deal of attention has been paid to a remarkable illness. Psychologists and psychiatrists term this "pathological questioning or doubt" [*Grübelsucht; Zweifelsucht*]; it would perhaps better be termed "pathological skepticism." One now encounters innumerable instances of this illness in the most remarkable forms, and it is already necessary that the study of this disease in particular be promoted within the cultural context of our time. This illness manifests itself—you can learn a great deal about it from the psychiatric literature—in these people, from a certain age onward, usually from puberty or the period immediately preceding puberty, no longer being able to relate properly to the external world. When confronted with their experiences in the external world, these people are overcome by an infinite number of questions. There are certain individuals who, though they remain otherwise fully rational, can pursue their duties to a great extent and are fully cognizant of their condition, must begin to pose the most extraordinary questions if they are but slightly withdrawn from what normally binds them to the external world. These questions simply intrude into their life and cannot be brushed aside. They intrude themselves especially strongly in individuals with healthy, or even conspicuously healthy, organizations —in individuals who have an open mind and a certain understanding for the manner in which modern scientific thinking proceeds. They experience modern science in this way, so that they cannot understand at all how such questions arise unconsciously thereby. Such phenomena are evident especially in women, who have less robust natures than men and who also tend to acquire their knowledge of natural science, if they undertake to do so, not so much through the highly disciplined scientific literature but rather through works intended for laymen and *dilettanti*. For if at this time immediately before puberty, or just when puberty is on the wane, there should occur an intense preoccupation with modern scientific thought in the way I have just described, among such people a high incidence of this disease can be observed. It manifests itself in these people having then to ask: where ever does the sun come from? And no matter how clever the answers one gives them, one question always calls forth another. Where does the human heart come from? Why does it beat? Did I not forget two or three sins at confession? What happened when I took Communion? Did a few crumbs of the Host perhaps fall to the ground? Did I not try to mail a letter somewhere and miss the slot? I could produce a whole litany of such examples for you, and you would see that all this is eminently suited to keeping one uneasy.

Now, when the spiritual scientist comes to consider this matter he feels himself right at home. It is simply a manifestation of the element in which the spiritual scientist resides consciously when he achieves an experience of the toneless-musical speech of spiritual beings through Inspiration. Those afflicted with pathological skepticism enter this region unconsciously. They have cultivated nothing that would enable them to comprehend the state into which they enter. The spiritual scientist knows that throughout the entire night, from falling asleep until waking, one lives in an element consisting entirely of such questions, that out of the sleeping state countless questions arise within one. The spiritual scientist knows this condition, because he can experience it consciously. Whoever approaches these matters from the standpoint of normal consciousness and seeks thus to comprehend them will perhaps make attempts at all kinds of rationalistic explanations, but he will not arrive at the truth, because he is unable to comprehend the matter through Inspirative cognition. Such a one sees that there are, for example, people who go to the theater in the evening and on leaving the theater are helpless to resist the countless questions that overcome them: what is this actress's relationship to the outer world? What was that actor doing some previous year? What are the relationships between the individual actors and actresses? How was this or that flat constructed? Which painter is responsible for each? and so on, and so on. For days on end such people are subject to the influence of this pesky questioner within. This is a pathological condition that one begins to understand only by realizing that these people enter a region the spiritual scientist experiences in Inspiration by approaching this realm differently from those afflicted with this pathological condition. Persons in this pathological state enter the same region as the spiritual scientist, but they do not take their egos with them; in a certain sense they lose their egos upon entering this realm. And it is just this ego that is the ordering faculty. It is the ego that is capable of bringing the same kind of order into this world as we are able to bring to our physical environment. The spiritual scientist knows that one lives in this same region between falling asleep and waking. Everyone who returns from the theater actually is deluged by all these questions in the night while he sleeps, but due to the operation of certain laws sleep normally spreads itself out over this interlocutor, so that one has finished with him by the time one awakes again.

In order to perform valid spiritual research, one must bear into this region unimpaired judgment, complete discretion, and the full force of the human ego. Then we do not live in this region in a kind of super-skepticism but rather with just as much self-possession and confidence as in the physical world. And actually all the meditative exercises that I have given in my book, *Knowledge of the Higher Worlds and its Attainment*, are intended in large part to result in a greater ability to enter this region preserving one's ego in full consciousness and in strict inner discipline. The purpose of a large part of the spiritual scientist's initial schooling is to keep him from losing the inner support and discipline of the ego while traversing this path.

The finest example in recent times of a man who entered this region without full preparation is someone whom Dr. Husemann has characterized here in another context. The finest example is Friedrich Nietzsche. Nietzsche is, to be sure, an extraordinary personality. In a certain sense he was not an intellectual at all. He was not your conventional scholar. With the tremendous gifts of genius, however, he grew out of puberty into scientific research; with these tremendous gifts he was able to take in what the contemporary sciences can offer. That, despite having acquired this knowledge, he did not become a scholar of the conventional sort is shown quite simply by the polemics of so exemplary a modern scholar as Wilamowitz, who came out in opposition immediately after the appearance of the young Nietzsche's

65

first publication. Nietzsche had just published his treatise, *The Birth of Tragedy out of the Spirit of Music*, in which there resounds a readiness to undergo initiation, to enter the musical, the Inspirative—even the title reveals his yearning for the realm that I have characterized—but he could not. The possibility did not exist. In Nietzsche's time a conscious spiritual science did not exist, but in giving his work the title, *The Birth of Tragedy out of the Spirit of Music*, he indicated that he wished to come to terms with a phenomenon such as Wagnerian tragedy out of this spirit of music. And he entered further and further into this realm. As I said, Wilamowitz immediately came out in opposition and wrote his polemics against *The Birth of Tragedy*, in which he completely rejected from his academic point of view what Nietzsche, unschooled but yearning for knowledge, had written. From the point of view of modern science he was of course completely justified. And actually it is hard to understand how so excellent a thinker as Erwin Rohde could have believed a compromise was possible between this modern philology that Wilamowitz represented and what lived within Nietzsche as a dark striving, as a yearning for initiation, for Inspiration. What Nietzsche had acquired in this manner, had inwardly appropriated, grew out into the other fields of contemporary sciences. It grew into positivism, namely that of the Frenchman, Comte, and the German, Dühring. While cataloguing Nietzsche's library in the 1890s I saw with my own eyes all the marks Nietzsche had so conscientiously made in the margins of Dühring's works, from which he acquired his knowledge of positivism; I held all these books in my own hand. I could enter sympathetically right into the manner in which Nietzsche took positivism up into himself. I could well imagine how he then reverted to an extra-corporeal existence, where he experienced this positivism again without having penetrated into this region sufficiently with his ego. As a result, he produced works such as *Human, All Too Human*, exhibiting a constant oscillation between an inability to move within the world of Inspiration and a desire to remain there nonetheless. One notices this in the aphoristic progression of Nietzsche's style in these works. Nietzsche strives to bring his ego into this realm, but it tears itself away again and again: thus he produces not a systematic, artistic presentation but only aphorisms. It is just this constant self-interruption in aphorism that reveals the inward soul of this remarkable spirit. And then he rises to encounter that which has provided modern science, the contemporary physical sciences, with their greatest riddles. He rises up to encounter what lives in Darwinism, what lives in the theory of evolution, and attempts to demonstrate how the most complicated organisms have gradually arisen out of the most primitive. He penetrates into this realm, a realm into which I have sought in a modest way to bring inner structure and an inward mobility—you can follow this in the discussion of Haeckel in my book, *The Riddles of Philosophy*. Nietzsche enters this realm, and there emerges from his soul the notion of a kind of super-evolution [*Überevolutionsgedanke*]. He follows the course of evolution up to man, where this notion of evolution explodes to create his "superman." In following this self-progression of evolving beings he loses the content, because he is unable to obtain the true content through Inspiration: he is confined to the empty idea of "eternal recurrence."

Only by virtue of the inner integrity of his personality was Nietzsche able to avoid what the pathologist calls "pathological skepticism." It was something within Nietzsche, a prodigious health that Nietzsche himself sensed underlying his debility, that asserted itself and kept him from falling prey to complete skepticism, leading him rather to contrive what later became the content of his most inspir-

67

ing works. No wonder, then, that this excursion into the spiritual world, this striving to proceed from music to the inner word, to inner being, culminated in the most unmusical of ideas—that of "the eternal recurrence of the same"—and the empty, merely lyrical "superman." No wonder that it had to end in the condition that his physician, for example, diagnosed as an "atypical case of paralysis."

Yes, this man who did not know Nietzsche's inner life, who was incapable of judging it from the standpoint of spiritual science and confronted the images and ideas of Nietzsche's inner life as a mere psychiatrist, without sympathetic understanding—this man found only an abstraction to answer the question posed by the concrete case before him. With regard to all nature Du Bois-Reymond had said in 1872: *ignorabimus*. Confronted with exceptional cases, the psychiatrist says: paralysis, atypical paralysis. Confronted with concrete cases that reveal the essence of present human evolution, the psychiatrist can say only *ignorabimus*, or *ignoramus*. This is but a translation of what is clothed in the words "atypical case of paralysis."

This eventually destroyed Nietzsche's body. It produced the condition that makes Nietzsche such a revealing phenomenon within our contemporary cultural life. This is the other form of the debility appearing in certain highly cultivated individuals, which psychiatrists term pathological doubt or hyperskepticism. And the phenomenon of Nietzsche—here I must be allowed a personal remark—stood before my eyes the moment that, trembling, I entered his room in Naumburg a few years after his illness. He lay upon the sofa after dinner, staring into space. He recognized nobody around him and stared at one like a complete idiot, but the light of his former genius still gleamed within his eyes.

If one looked at Nietzsche knowing all one could about

68

his world view, about the ideas and images that lived within his soul; if, unlike the mere psychiatrist, one stood before Nietzsche, this ruin of a man, this physical wreck, with this image in one's soul, then one knew: this man strove to view the world revealed by Inspiration. Nothing of this world came forth to him. And the part of him that desired to achieve Inspiration finally extinguished itself: for years the physical organism was filled by a soul-spirit devoid of content.

From such a sight one can learn the whole tragedy of our modern culture, its striving for the spiritual world, its inclination toward that which can proceed from Inspiration. For me—and I do not hesitate in the slightest to introduce a personal remark here—this was one of those moments that can be interpreted in a Goethean manner. Goethe says that nature conceals no secret that she is not willing to reveal in one place or another. No, the entire world contains not a single secret that is not revealed in one place or another. The present stage of human evolution conceals the secret that humanity is giving birth to a striving, an inclination, an impulse that rumbles within the social upheavals our civilization is undergoing—an impulse that seeks to view the spiritual world of Inspiration. And Nietzsche was the one point where nature revealed its open secret, where the striving that exists within humanity as a whole could reveal itself. We must seek this if all those striving for education, seeking within modern science—and this shall be the entire civilized world, for educaton must become universal—if humanity as a whole is not to lose its ego and civilization fall into barbarism.

That is one great cultural anxiety, one great threat to civilization, which must be faced by anyone who follows the contemporary progress of human evolution and seeks to develop a thinking that can grasp the realities of social life. Similar phenomena assert themselves on the other side as well, on the side of consciousness. And we shall have to study these phenomena on the side of consciousness at least in outline as well. We shall see how these other phenomena arise out of the chaos of contemporary life, phenomena that appear pathologically and have been described by Westphal, Falret, and others. It is no accident that these have been described only just in the most recent decades. On the other side, that of the boundary of consciousness, we encounter the phenomena of claustrophobia, astraphobia, and agoraphobia[6], just as we encounter pathological skepticism on the side of matter. And in the same way (we shall discuss this further) in which pathological skepticism must be cured culturally-historically through the cultivation of Inspiration—one of the great tasks of contemporary social ethics—we are threatened with the emergence of the phenomena that I shall describe tomorrow: claustrophobia, astraphobia, and agoraphobia. These emerge pathologically and can be overcome through Imagination, which, when civilization has acquired it, shall become a social blessing for all humanity.

VI

Yesterday I closed with a consideration of what reveals itself at one boundary of scientific thinking as a real and true mode of cognition: I closed with a characterization of Inspiration. I have brought to your attention the way in which man enters through Inspiration a spiritual world: he knows that he is in this world and feels also that he is outside the body. I have shown you how the transition from the experience of a "toneless" musical element to a merger with an individuated element of being occurs. It also became clear in the course of yesterday's considerations of pathological skepticism and hypercriticism that pathological conditions can arise within man if he takes this step out of the body without the accompaniment of the ego, if he does not suffuse the conditions he experiences in Inspiration with full self-consciousness. If one brings the ego into Inspiration, Inspiration represents a healthy, indeed a necessary, step forward in human cognition. Yet in a cultural epoch such as ours, in which man's being is striving to free itself from the physical organism, one cannot allow this condition to come about in an instinctive, unconscious, unhealthy way without the emergence of the pathological conditions we discussed yesterday. For, you see, there exist two poles in human nature. We can either turn to what opens a free, spiritual vision of the highest realities, or, by shunning this, by not summoning sufficient courage to penetrate into these regions with full consciousness but allowing ourselves to be

71

driven by unconscious forces within ourselves, we can call forth illness in the physical organism. And it would be a grave error to believe that one could guard against this illness by electing not to strive into the actual spiritual world. Illness will occur anyway, if the instincts are allowed to drive the astral body, as we call it, out of the organism. Yet especially at the present time, even if we do not investigate the spiritual world ourselves, we are fully protected against the pathological states that I described yesterday—even against those arising only in the soul—by seeking to comprehend rationally the ideas of spiritual science.

What is it, however, that we bear into the spiritual world when we take full consciousness with us? You need only follow somewhat man's development from birth to the change of teeth and beyond in order to realize that, besides the development of speech, thinking, and so forth, an especially important element in this human development is the gradual emergence and transformation of memory. If you then look at the course of human life, you will come to see the tremendous importance of memory for a fully human existence. If, as a result of certain pathological conditions, the continuity of memory is interrupted, so that we cannot recall certain experiences we have had, then a serious illness befalls us, for we feel that the thread of the ego, which otherwise runs through our lives, has been broken. You can consult my book, *Theosophy*,[7] on this: memory is intimately connected with the ego. Thus in pursuing the path I have characterized we must take care not to lose what manifests itself in memory. We must take along with us into the world of Inspiration the power of soul that provides us with memory.

Just as in nature everything changes, however—just as the plant, in growing, metamorphoses its green leaves into the red petals of the flower; just as everything in nature is in constant metamorphosis, so it is with everything concerning human existence. If we really bear the faculty of memory out into the world of Inspiration under the full influence of ego-consciousness, it metamorphoses itself. Then one comes to realize that in the moment of one's life in which one investigates the spiritual world in Inspiration, one does not have the normal faculty of memory at one's disposal. One has this faculty of memory at one's disposal in healthy life within the body; outside the body, this faculty is no longer available.

This results in something extraordinary—something that, since I present it to your mind's eye for the first time, might seem paradoxical, yet that is fully grounded in reality. Whoever has become a true spiritual scientist, who enters and seeks to experience through Inspiration actual spiritual reality as I have described it in my books, must experience this reality each time anew if he wishes to have it present to consciousness. Thus whenever someone speaks out of Inspiration concerning the spiritual world—not from notes or from mere memory but when he expresses immediately what reveals itself to him in the spiritual world—he must perform the task of spiritual perception each time anew. The faculty of memory has transformed itself. One has retained only the power to call forth the experience again and again. For that reason the spiritual scientist does not have it so easy as one who relies on mere memory. He cannot simply communicate some information out of memory but must call forth anew each time what presents itself to him in Inspiration. In this matter it is essentially the same as it is in normal sense perception of the physical world. If you wish actually to perceive within the physical world of the senses, you cannot turn away from what you wish to perceive and still have the same perception in another place. You must return to the object. In the same way, the spiritual scientist must return to the same spiritual content of consciousness. And just as in physical perception one must learn to move about in space in order to perceive this or that in turn, the spiritual scientist who has attained Inspiration must learn to move freely within the element of time. He must be able—if you will allow me to use a paradoxical expression—to swim within the element of time. He must learn to travel along with time itself, and when he has learned this, he finds that the faculty of memory has undergone a metamorphosis, that the faculty of memory has transformed itself into something else. What memory performed within the physical world of the senses must be replaced by spiritual perception. This transformed memory, however, gives the spiritual scientist perception of a more encompassing ego. Now the ego is recognized to be more encompassing. When one has transformed memory, which contains the power of the ego between birth and death, the content of the ego cracks the husk that circumscribes but *one* lifetime. Then the fact of repeated earthly incarnations, alternating with a purely spiritual existence between death and rebirth, emerges as something that can be grasped as a reality.

On the other side, the side of consciousness, there emerges something different when one seeks to avoid what an ancient view of the spirit, that of the Vedanta, did not yet know. We in the West feel on the one hand the loftiness of the spiritual view when we steep ourselves in the ancient Oriental wisdom. We feel that in the Vedanta the soul was borne up into spiritual regions in which it could move in a way that the Westerner's normal consciousness can only in mathematical, geometrical, analytic-mechanical thinking. When we descend into the expansive realms that in the Orient were accessible to normal consciousness, however, we find something that we Westerners, because of our more advanced state of evolution, can no longer bear: we find an extensive symbolism, an allegorization of the natural world. It is this symbolism, this allegorization, this thinking about external nature in images, that makes us clearly aware that we are being led away from reality, away from a true investigation of nature. This has become part of certain religious confessions. Certain religious confessions are at a loss how to proceed with this act of symbolization, of mythologization, which has become decadent. For us in the West, that which the Oriental, living in an illusory world, applied directly in this way to external nature, that with which he believed himself capable of arriving at insights concerning the natural world—for us at present this has value only as an exercise preliminary to further spiritual research. We must acquire the soul faculty that the Oriental employed in symbolism and anthropomorphism. We must exercise this faculty inwardly and remain fully conscious thereby: we lapse into superstitions, into rhapsodic enthusiasm for nature, if we employ this faculty to any end but the cultivation of our soul. Later I shall have occasion to speak here about the particulars of this—which, by the way, you can find in my book, *Knowledge of the Higher Worlds and its Attainment.*

By taking this faculty that the Oriental turned outward and employing it inwardly, as an activity of inner schooling [*Kraft des Übens*], by first developing a pictorial representation in such a way within, one actually begins to arrive at new insights on the other side, on the side of consciousness. One gradually achieves a transformation of abstract, merely notional thinking into pictorial thinking. Then there arises what I can only call an experiential thinking [*erlebendes Denken*]. One experiences pictorial thinking. Why does one experience this? One experiences nothing other than what is active within the physical body during the first years of childhood, as I have described it to you. One experiences not the human organism that has taken static form in space but rather what lives and weaves within man. One experiences it in pictures. One gradually struggles through to a viewing of the life of the soul in its actuality. On the other side the content of consciousness gradually emerges within cognition: pictorial representation, a life within Imaginations. And without entering into this life of Imaginations, modern psychology shall not progress. In this way, and in this way only, by entering into Imagination, there will arise again a psychology that is more than word-games, a psychology that actually looks into the soul of man.

Just as the time has come in which, as a result of general cultural relationships, man is gradually excarnating from the physical body and striving for Inspiration, as we have seen in the example of Nietzsche, the time has come in which man, if he desires self-knowledge, should feel himself led toward Imagination. Man must descend deeper into himself than was necessary in the course of previous cultural history. If evolution is not to lapse into barbarism, humanity must attain a true image of itself [*Selbstschau*], and humanity can accomplish this only by accepting the knowledge offered by Imagination. That man is striving to descend deeper into his inner self than has been the case in evolution heretofore is shown, again, in the phenomena of pathological diseases of a particularly modern form. These have been described very recently by those who are able to study them from the point of view of medicine or psychiatry. It is shown above all in the emergence of agoraphobia, claustrophobia, and astraphobia—illnesses of a sort that arise especially frequently in our time. Even if they usually are observed only as pathological conditions requiring psychiatric treatment, the more acute observer can see something else altogether. He sees agoraphobia, astraphobia, and so forth already emerging from the soul-nature of humanity, just as he saw Inspiration arising pathologically in Friedrich Nietzsche. Above all, he can observe states of soul that often appear outwardly normal from which emerges agoraphobia—morbid dread of open spaces. He sees emerging something that appears as astraphobia, a state in which one fails to come to terms with an inner sensation. This inner feeling can grow to the extent that the organs of digestion are attacked, and digestion is disturbed. He comes to know what might be called fear of isolation, agoraphobia[8], in which one cannot remain alone but only where there is company assembled all around and so forth. Such things emerge. These things show that humanity is presently striving for Imagination and that an illness that must otherwise become an illness of the entire culture can be counteracted only by developing Imagination. Agoraphobia—this is an illness that manifests itself in many people in a frightening way. These people grow up, and from a certain point in their lives onward remarkable conditions manifest themselves. If such a person steps out of the house into a square devoid of people he is stricken with a fear that is entirely incomprehensible to him. He is afraid of something; he does not dare go a step further into the empty square, and if he does, it can happen that he falls down on his knees or perhaps even topples over in a faint. The moment that even a child comes, the sufferer grasps its arm or merely reaches out to touch the child: in this moment he feels himself inwardly strengthened again, and the agoraphobia subsides. One case that has been described in the medical literature is particularly interesting. A young man who felt himself strong enough even to become an officer is overcome by agoraphobia while on maneuvers as he is sent out to map some terrain. His fingers tremble; he is unable to draw. Wherever there is emptiness around him, or what he perceives as emptiness, he is beset with fears that he immediately senses to be pathological. He is in the vicinity of a mill. In order to be able to perform his duty at all, he must keep a small child at his side, and its mere presence is enough for him to be able to resume drawing. We ask ourselves: what is the cause of such phenomena? Why is it that there are, for example, people who, when they have somehow forgotten to leave open the door to their bedrooms at night—something that has perhaps long since become a habit with them —wake in the night dripping sweat and can do nothing but leap up to open the door, for they cannot stand to be in an enclosed space. There are such people. Some suffer to such an extent that they must have all the doors and windows open. If their house is on a square, they must leave open the door leading out, so that they know they are free and can get out into the open at any time. This claustrophobia is something that one sees emerging—even if it often does not emerge in so radical a form—if one is able to observe human states of soul more closely.

And then there are people who feel, even to a physical degree, something inexplicable happening within them. What is it? It is an approaching thunderstorm or some other atmospheric condition. There are otherwise intelligent people who must draw the curtains whenever there is lightning or thunder. Then they must sit in a dark room, for only in this way can they protect themselves from what they experience in the atmospheric conditions. This is astraphobia, or morbid fear of thunderstorms. What is the cause of these states that we observe already very clearly in the souls of human beings today, especially in those who for a long time surrender themselves devoutly to a certain dogmatism? In these people one observes precisely these states of soul, even if they have not manifested themselves yet physically. These states are just beginning to appear. Their emergence works to upset a balanced, calm approach to life. They also emerge in such a way that they call forth all kinds of pathological conditions that are ascribed to every sort of thing, because the physical symptoms of claustrophobia, agoraphobia, or astraphobia are not yet manifest, while they must actually be ascribed to the particular configuration of soul arising within man.

What is the cause of such conditions? They are the result of our need not only to experience the life of the soul discarnately but also to bring this experience of the discarnate soul down into the physical body. We must allow it to immerse itself consciously. Just as that which I have described to you in the course of these lectures gradually extricates itself from the body between birth and the change of teeth, so also that which is experienced externally, which we could call experience of the astral, immerses itself again in the physical organism between the change of teeth and puberty. And what takes place in puberty is nothing other than this immersion between approximately the seventh and fourteenth years. The independent soul-spirit that man has developed must immerse itself in the body again, and what then emerges as physical love, as sexual desire, is nothing other than the result of this immersion I have described to you. One must come to understand this immersion clearly. Whoever wishes to gain a true understanding of the basis of consciousness must be able to effect this in a fully conscious, healthy way, using such methods as I shall describe here later. That is to say, he must learn to immerse himself in the physical body. Then he attains an initial experience of what manifests itself as an Imaginative representation of the inner realm. Here a faculty of formal representation framed for an external, three-dimensional world of plastic forms is insufficient. To perform this inner activity one needs a mobile faculty of formal representation: one must be able to overcome gradually everything spatial in Imagination and to immerse oneself in the representation of something intensive, something that radiates activity. In short, one must immerse oneself in such a way that in descending one can still clearly differentiate between oneself and one's body. Whatever inheres in the subject cannot be known. If one can keep what one experiences outside from immersing unconsciously in the physical body, one descends into the physical body and experiences in descending the essence of this body up to the level of consciousness in Imagination, in pictures.

Whoever fails to keep these pictures separate, however, and allows them to slip into the physical body, confronting the physical body not as an object but as something subjective, brings the sensation of space down into the physical body with him. The astral thereby coalesces with the physical to a greater degree than should be allowed. The experience of the external world coalesces with man's inner life, and because he makes subjective what should have remained objective, he can no longer experience space normally. Fear of empty space, fear of lonely places, fear of the astrality diffused through space, of storms, perhaps even of the moon and stars, rise up within one. One lives too deeply within oneself. Thus it is necessary that all exercises leading to the life of Imagination protect one against descending too deeply into the body. One must immerse oneself in the body in such a way that the ego remains outside. One may not take the ego out into the world of Imagination in the way that one must carry the ego out into the world of Inspiration. Although one worked toward Imagination through a process of symbolization, through pictorial representation, in Imagination itself all pictures created by mere fantasy disappear. Now objective pictures emerge instead. Only that which actually lives within the human form ceases to confront one as an object. One loses the outward human form

80

and there emerges a diversity of living forms from the human etheric. One now sees not the unified human form but the profusion of animal forms that interpenetrate and merge to create the human form. One comes to know in an inward way what lives within the realms of plants and minerals. One learns this through introspection. One learns what can never be learned through atomism and molecularism: one learns what actually lives within the realms of plants and minerals. And how is it that we avoid bringing the ego down into the physical body when we strive for Imagination? Only by developing the power of love more nobly than in normal life, where love is led by the powers of the bodily senses. Only by acquiring the selfless power of love, freedom from egotism not only regarding the realm of humanity but also regarding the realm of nature. Only by allowing all that leads to Imagination to be borne by love, by merging this power of love with every object of cognition that we seek in this manner.

Again we have divergent tendencies: the healthy tendency to extend the power of love into Imagination or the pathological tendency to expose ourselves to fear of what is outside. We experience what lies outside with our ego and then, without restraining our ego, bear it down into the body, giving rise to agoraphobia, claustrophobia, and astraphobia. Yet we enjoy the prospect of an extremely high mode of cognition if we can develop in a healthy way what threatens humanity in its pathological form and would lead it into barbarism.

In this way one attains a true knowledge of man. One surpasses all that anatomy, physiology, and biology can teach; one attains a true knowledge of man by actually seeing through the physical body. Oh, man comes to know himself in a way so different from that which nebulous mystics believe, who think that some abstract divinity

81

reveals itself to them when they delve down within. Oh no, something rich and concrete reveals itself; something that provides insight into the human organism, into the nature of the lungs, the liver, and so forth. Only this can be the basis of a true anatomy, a true physiology; only this can serve as the basis for a true understanding of man and also for a true medical science. One has developed two faculties within human nature. On the side of matter is the faculty of Inspiration, developed by gradually discovering within matter a spiritual realm that expands out into the tableau Mr. Arenson has depicted for you here. The other faculty is developed by discovering within oneself the realms that I described as the basis of a true knowledge of man, of a true medical science, when I spoke here earlier this year before almost forty medical doctors.

These two faculties, however, those of Inspiration and Imagination, can join together. The one can coalesce with the other, but it must happen in full consciousness and by comprehending the cosmos in love. Then there arises a third faculty, a confluence of Imagination and Inspiration in true, spiritual Intuition. Then we rise up to that which allows us to recognize the external material world to be a spiritual world, the inner realm of the soul and spirit with its material foundations as a continuous whole; we rise up to that which grants us knowledge of the expansion of human existence beyond earthly life, as I have described it to you here in other lectures. One comes thus on the one side to know the realms of plants, animals, and minerals in their inmost essences, in their spiritual content, through Inspiration. By coming to know the human organs through Imagination one creates the basis for a true organology, and by uniting in Intuition what one has learned about plants, animals, and minerals with what Imagination reveals concerning the human organs, one attains a true therapy, a science of medication that knows in a real sense how to apply the external to the internal. The true doctor must understand medications cosmologically; he must understand the human organs anthropologically, or actually anthroposophically. He must come to grasp the external world through Inspiration, the inner world through Imagination, and he must achieve a therapy based upon real Intuition.

You see what a prospect opens before us if we are able to comprehend spiritual science in its true form. To be sure, this spiritual science still has to shed many externals and much that still adheres to it in the minds of those who believe they can nurture it with fantasies and dilettantism of every sort. Spiritual science must develop a method of research as rigorous as mathematics and analytical mechanics. On the other hand, spiritual science must rid itself of all superstitions. Spiritual science must truly be able to call forth in light-filled clarity the love that otherwise overcomes man if he can call it forth out of instinct. Then spiritual science will be a seed that will grow and send its forces out into all the sciences and thus into human life.

For this reason, let me bring to a close what I have had to say to you in these lectures with one more brief consideration. Beforehand I would like to say that there is, of course, still much that can be read between the lines of my descriptions. Some of this I shall make legible in two lectures this evening and tomorrow: they will elaborate what I could only intimate in the short time available to us for this course. Only what is gained by attaining Imagination on the one hand and Inspiration on the other, and then uniting Imagination and Inspiration in Intuition, gives man the inner freedom and strength enabling him to conceive ideas that can then be effected in social life. And only those who experience contemporary life with a sleeping soul can fail to see everything that is brewing in the most frightful way, threatening a horrific future.

What is the spiritual cause of this? The spiritual cause of this is something one can perceive by studying attentively recent human evolution as it manifests itself in extremely prominent individuals. How human beings strove in the later nineteenth and early twentieth centuries to arrive at clear concepts, to arrive at truly inward, clear impulses for three concepts that are of the very greatest importance for social life: the concept of capital, the concept of labor, and the concept of commodities! Just look at the relevant literature from the nineteenth and the beginning of the twentieth centuries to see how human beings strove to understand what capital actually means within the social process, to see how that which human beings strove to understand in concepts has passed over into frightful struggles in the external world. Just look how intimately the particular feeling emerging within humanity in the present age corresponds to what they are able to feel and think concerning the function, the meaning of labor within the social organism. Then look at the hopelessly inadequate definition of "commodity"! Human beings strove to bring three practical concepts into clear focus. In the course of life in the civilized world today one sees everywhere a lack of clarity regarding the triad, capital, labor, commodities. And one cannot rise up to answer the question: what function does capital have within the social organism? One is able to answer this question only when, out of a true spiritual science, by means of Imagination and Inspiration united in Intuition, one understands that a proper impulse for the functioning of capital can be found within the spiritual life as an independently subsisting part of the social organism. Only true Imagination can bring real comprehension of this part of the social organism. And one will come to realize something else as well. One will realize that one can come to understand labor's functioning within the social organism when one no longer understands what is produced by human labor in terms of the product, so that one no longer conceives commodities in the Marxist manner as congealed labor or even congealed time. Rather, one will realize that the results of human labor can be understood by arriving at a representation, at a free experience of that which can proceed from man. The concept of labor will become clear only to those who know what is revealed to man through Inspiration.

And the concept of "commodity" is the most complicated imaginable. For no single man is able to comprehend what commodities are in their actual existence in life. Anyone who wishes to define commodities has not the slightest inkling what knowledge is. "Commodity" cannot be defined, for one can define in this sense or formulate conceptually only what concerns but one individual, what one man alone can comprehend with his soul. Commodities, however, always exist in the interaction between a number of human beings and a number of individuals of a certain type. Commodities exist in the interaction between producers, consumers, and those who mediate between them. The impoverished concepts of barter and purchase, products of a discipline that fails to recognize the limits of natural science, shall never prove adequate to an understanding of commodities. Commodities, the products of human labor, exist in the relationship between several individuals, and if a solitary man undertakes to understand commodities "as such," he is on the wrong track. Commodities must be understood as a function of the socially contracted majority of human beings, of association. Commodities must be understood in terms of association; they must exist in association. Only when associations are formed that process what originates with the producers, businessmen, and consumers will there arise—not out of the individual but through association, through the worker associations—the social concept, the concept of "commodity," that human beings must share before there can exist a healthy economic life.

If human beings would only take the trouble to ascend to that which the spiritual scientist can convey from the realm of higher cognition, they would find concepts giving rise to the social forms we must develop if we wish to reverse the course of a civilization on the decline. It is thus no mere theoretical interest, no mere scientific need, that underlies all we shall strive for here. It is rather the most urgent need that the work and the research we do here make human beings mature enough that they can go forth from this place to all the corners of the earth, taking with them such ideas and social impulses as really can buoy up an age so rapidly sinking and reverse the course of a world so clearly in decline.

VII

It is to be hoped that my discussions of the boundaries of natural science have been able to furnish at least some indications of the difference between what spiritual science calls knowledge of the higher worlds and the mode of knowledge proceeding from everyday consciousness or ordinary science. In everyday life and in ordinary science our powers of cognition are those we have acquired through the conventional education that carries us up to a certain stage in life and whatever this education has enabled us to make of inherited and universally human qualities. The mode of cognition that anthroposophically oriented spiritual science terms knowledge of the higher worlds has its basis in a further self-cultivation, a further self-development; one must become aware that in the later stages of life one can advance through self-education to a higher consciousness, just as a child can advance to the stage of ordinary consciousness. The things we sought in vain at the two boundaries of natural science, the boundaries of matter and of ordinary consciousness, reveal themselves only when one attains this higher consciousness. In ancient times the Eastern sages spoke of such an enhanced consciousness that renders accessible to man a level of reality higher than that of everyday life; they strove to achieve a higher development, similar to the one we have described, by means of an inner self-cultivation that corresponded to their racial characteristics and evolutionary stage. The meaning of what radiates forth from

87

the ancient Eastern wisdom-literature becomes fully apparent only when one realizes what such a higher level of development reveals to man. If one were to characterize the path of development these sages followed, one would have to describe it as a path of Inspiration. For in that epoch humanity had a kind of natural propensity to Inspiration, and in order to understand these paths into the higher realms of cognition, it will be useful if first we can gain clarity concerning the path of development followed by these ancient Eastern sages. I want to make it clear from the start, however, that this path can no longer be that of our Western civilization, for humanity is in a process of constant evolution, ever moving forward. And whoever desires—as many have—to return to the instructions given in the ancient Eastern wisdom-literature in order to enter upon the paths of higher development actually desires to turn back the tide of human evolution or shows that he has no real understanding of human progress. In ordinary consciousness we reside within our thought life, our life of feeling, and our life of will, and we initially substantiate what surges within the soul as thought, feeling, and will in the act of cognition. And it is in the interaction with percepts of the external world, with physical-sensory perceptions, that our consciousness first fully awakens.

It is necessary to realize that the Eastern sages, the so-called initiates of the East, cultivated perception, thinking, feeling, and willing in a way different from their cultivation in everyday life. We can attain an understanding of this path of development leading into the higher worlds when we consider the following. In certain ages of life we develop what we call the soul-spirit toward a greater freedom, a greater independence. We have been able to show how the soul-spirit, which functions in the earliest years of childhood to organize the physical body, emancipates itself, becomes free in a sense with the change of teeth. We have shown how man then lives freely with his ego in this soul-spirit, which now places itself at his disposal, while formerly it occupied itself—if I may express myself thus—with the organization of the physical body. As we enter into ever-greater participation in everyday life, however, there arises something that initially prevents this emancipated soul-spirit from growing into the spiritual world in normal consciousness. As human beings, we must traverse the path that leads us into the external world with the requisite faculties during our life between birth and death. We must acquire such faculties as allow us to orient ourselves within the external, physical-sensory world. We must also develop such faculties as allow us to become useful members of the social community we form with other human beings.

What arises is threefold. These three things bring us into a proper relationship with other human beings in our environment and govern our interaction with them. These are: language, the ability to understand the thoughts of our fellow men, and the acquisition of an understanding, or even a kind of perception, of another's ego. At first glance these three things—perception of language, perception of thoughts, and perception of the ego—appear simple, but for one who seeks knowledge earnestly and conscientiously these things are not so simple at all. Normally we speak of five senses only, to which recent physiological research adds a few inner senses. Within conventional science it is thus impossible to find a complete, systematic account of the senses. I will want to speak to you on this subject at some later time. Today I want only to say that it is an illusion to believe that linguistic comprehension is implicit in the sense of hearing, of that which contemporary physiology dreams to be the organization of the sense of hearing. Just as we have a sense of hearing, so also do we have a sense of lan-

89

guage. By this I do not mean the sense that guides us in speaking—for this is also called a sense—but that which enables us to comprehend the perception of speech-sounds, just as the auditory senses enable us to perceive tones as such. And when we have a comprehensive physiology, it will be known that this sense of speech is analogous to the other and can rightfully be called a sense in and of itself. It is only that this sense extends over a larger part of the human constitution than the other, more localized senses. Yet it is a sense that nevertheless can be sharply delineated. And we have, in fact, a further sense that extends throughout virtually all of our body—the sense that perceives the thoughts of others. For what we perceive as word is not yet thought. We require other organs, a sensory organization different from that which perceives only words as such, if we want to understand within the word the thought that another wishes to communicate.

In addition, we are equipped with an analogous sense extending throughout our entire bodily organization, which we can call the sense for the perception of another person's ego. In this regard even philosophy has reverted to childishness in recent times, for one can often hear it argued: we encounter another man; we know that a human has such and such a form. Since the being that we encounter is formed in the way we know ourselves to be formed, and since we know ourselves to be ego-bearers, we conclude through a kind of unconscious inference: aha, he bears an ego within as well. This directly contradicts the psychological reality. Every acute observer knows that it is not an inference by analogy but rather a direct perception that brings us awareness of another's ego. I think that a friend or associate of Husserl's school in Göttingen, Max Scheler, is the only philosopher actually to hit upon this direct perception of the ego. Thus we must differentiate three higher senses, so to

90

speak, above and beyond the ordinary human senses: the sense that perceives language, the sense that perceives thoughts, and the sense that perceives another's ego. These senses arise within the course of human development to the same extent that the soul-spirit gradually emancipates itself between birth and the change of teeth in the way I have described.

These three senses lead initially to interaction with the rest of humanity. In a certain way we are introduced into social life among other human beings by the possession of these three senses. The path one thus follows via these three senses, however, was followed in a different way by the ancients —especially the Indian sages—in order to attain higher knowledge. In striving for this goal of higher knowledge, the soul was not moved toward the words in such a way that one sought to arrive at an understanding of what the other was saying. The powers of the soul were not directed toward the thoughts of another person in such a way as to perceive them, nor toward the ego of another in such a way as to perceive it sympathetically. Such matters were left to everyday life. When the sage returned from his striving for higher cognition, from his sojourn in spiritual worlds to everyday life, he employed these three senses in the ordinary manner. When he wanted to exercise the method of higher cognition, however, he needed these senses in a different way. He did not allow the soul's forces to penetrate through the word while perceiving speech, in order to comprehend the other through his language; rather, he stopped short at the word itself. Nothing was sought behind the word; rather, the streaming life of the soul was sent out only as far as the word. He thereby achieved an intensified perception of the word, renouncing all attempts to understand anything more by means of it. He permeated the word with his entire life of soul, using the word or succession of words in such a way that he could enter completely into the inner life of the word. He formulated certain aphorisms, simple, dense aphorisms, and then strove to live within the sounds, the tones of the words. And he followed with his entire soul life the sound of the word that he vocalized. This practice then led to a cultivation of living within aphorisms, within the so-called "mantras." It is characteristic of mantric art, this living within aphorisms, that one does not comprehend the content of the words but rather experiences the aphorisms as something musical. One unites one's own soul forces with the aphorisms, so that one remains within the aphorisms and so that one strengthens through continual repetition and vocalization one's own power of soul living within the aphorisms. This art was gradually brought to a high state of development and transformed the soul faculty that we use to understand others through language into another. Through vocalization and repetition of the mantras there arose within the soul a power that led not to other human beings but into the spiritual world. And if, through these mantras, the soul has been schooled in such a way and to such an extent that one feels inwardly the weaving and streaming of this power of soul, which otherwise remains unconscious because all one's attention is directed toward understanding another through the word; if one has come so far as to feel such a power to be an actual force in the soul in the same way that muscular tension is experienced when one wishes to do something with one's arm, one has made oneself sufficiently mature to grasp what lies within the higher power of thought. In everyday life a man seeks to find his way to another via thought. With this power, however, he grasps the thought in an entirely different way. He grasps the weaving of thought in external reality, penetrates into the life of external reality, and lives into the higher realm that I have described to you as Inspiration.

Following this path, then, we approach not the ego of the other person but the egos of individual spiritual beings who surround us, just as we are surrounded by the entities of the sense world. What I depict here was self-evident to the ancient Eastern sage. In this way he wandered with his soul, as it were, upward toward the perception of a realm of spirit. He attained in the highest degree what can be called Inspiration, and his constitution was suited to this. He had no need to fear, as the Westerner might, that his ego might somehow become lost in this wandering out of the body. In later times, when, owing to the evolutionary advances made by humanity, a man might very easily pass out of his body into the outer world without his ego, precautionary measures were taken. Care was taken to ensure that whoever was to undergo this schooling leading to higher knowledge did not pass unaccompanied into the spiritual world and fall prey to the pathological skepticism of which I have spoken in these lectures. In the ancient East the racial constitution was such that this was nothing to fear. As humanity evolved further, however, this became a legitimate concern. Hence the precautionary measure strictly applied within the Eastern schools of wisdom: the neophyte was placed under an authority, but not any outward authority—fundamentally speaking, what we understand by "authority" first appeared in Western civilization. There was cultivated within the neophytes, through a process of natural adaptation to prevailing conditions, a dependence on a leader or guru. The neophyte simply perceived what the leader demonstrated, how the leader stood firmly within the spiritual world without falling prey to pathological skepticism or even inclining toward it. This perception fortified him to such an extent on his own entry into Inspiration that pathological skepticism could never assail him.

Even when the soul-spirit is consciously withdrawn from

the physical body, however, something else enters into consideration: one must re-establish the connection with the physical body in a more conscious manner. I said this morning that the pathological state must be avoided in which one descends only egotistically, and not lovingly, into the physical body, for this is to lay hold of the physical body in the wrong way. I described the natural process of laying hold of the physical body between the seventh and fourteenth years, which is through the love-instinct being impressed upon it. Yet even this natural process can take a pathological turn: in such cases there arise the harmful afflictions I described this morning as pathological states. Of course, this could have happened to the pupils of the ancient Eastern sages as well: when they were out of the body they might not have been able to bind the soul-spirit to the physical body again in the appropriate manner. One further precautionary measure thus was employed, one to which psychiatrists—some at any rate—have had recourse when seeking cures for patients suffering from agoraphobia or the like. They employed ablutions, cold baths. Expedients of an entirely physical nature have to be employed in such cases. And when you hear on the one hand that in the mysteries of the East—that is, the schools of initiation, the schools that led to Inspiration—the precautionary measure was taken of ensuring dependence on the guru, you hear on the other hand of the employment of all kinds of devices, of ablutions with cold water and the like. When human nature is understood in the way made possible by spiritual science, customs that otherwise remain rather enigmatic in these ancient mysteries become intelligible. One was protected against developing a false sense of spatiality resulting from an insufficient connection between the soul-spirit and the physical body. This could drive one into agoraphobia and the like or to

seek social intercourse with one's fellow men in an inappro-
priate way. This represents a danger, but one which can and
should—indeed must—be avoided in any training that leads
to higher cognition. It is a danger, because in following the
path I have described leading to Inspiration one bypasses in
a certain sense the path via language and thought to the ego
of one's fellow man. If one then quits the physical body in a
pathological manner—even if one is not attempting to attain
higher cognition but is lifted out of the body by a pathological condition—one can become unable to interact socially
with one's fellow men in the right way. Then precisely that
which arises in the usual, intended manner through properly regulated spiritual study can develop pathologically.
Such a person establishes a connection between his soul-
spirit and his physical body: by delving too deeply into it he
experiences his body so egotistically that he learns to hate
interaction with his fellow men and becomes antisocial. One
can often see the results of such a pathological condition
manifest themselves in the world in quite a frightening manner. I once met a man who was a remarkable example of
such a type: he came from a family that inclined by nature
toward a freeing of the soul-spirit from the physical body
and also contained certain personalities—I came to know one
of them extremely well—who sought a path into the spiritual worlds. One rather degenerate individual, however,
developed this tendency in an abnormal, pathological way
and finally arrived at the point where he would allow
nothing whatever from the external world to contact his
own body. Naturally he had to eat, but—we are speaking
here among adults—he washed himself with his own urine,
because he feared any water that came from the outside
world. But then again I would rather not describe all the
things he would do in order to isolate his body totally from

the external world and shun all society. He did these things because his soul-spirit was too deeply incarnated, too closely bound to the physical body.

It is entirely in keeping with the spirit of Goetheanism to bring together that which leads to the highest goal attainable by earthly man and that which leads to pathological depths. One needs only slight acquaintance with Goethe's theory of metamorphosis to realize this. Goethe seeks to understand how the individual organs, for example of the plant, develop out of each other, and in order to understand their metamorphosis he is particularly interested in observing the conditions that arise through the abnormal development of a leaf, a blossom, or the stamen. Goethe realizes that precisely by contemplating the pathological the essence of the healthy can be revealed to the perceptive observer. And one can follow the right path into the spiritual world only when one knows wherein the essence of human nature actually lies and in what diverse ways this complicated inner being can come to expression.

We see from something else as well that even in the later period the men of the East were predisposed by nature to come to a halt at the word. They did not penetrate the word with the forces of the soul but lived within the word. We see this, for example, in the teachings of the Buddha. One need only read these teachings with their many repetitions. I have known Westerners who treasured editions of the Buddha's teachings in which the numerous repetitions had been eliminated and the words of a sentence left to occur only once. Such people believed that through such a condensed version, in which everything occurs only once, they would gain a true understanding of what the Buddha had actually intended. From this it is clear that Western civilization has gradually lost all understanding of Eastern man. If we simply take the Buddha's teachings word for word; if we take the content of these teachings, the content that we, as human beings of the West, chiefly value, then we do not assimilate the essence of these teachings: that is possible only when we are carried along with the repetitions, when we live in the flow of the words, when we experience the strengthening of the soul's forces that is induced by the repetitions. Unless we acquire a faculty for experiencing something from the constant repetitions and the rhythmical recurrence of certain passages, we do not get to the heart of Buddhism's actual significance.

It is in this way that one must gain knowledge of the inner nature of Eastern culture. Without this acquaintance with the inner nature of Eastern culture one can never arrive at a real understanding of our Western religious creeds, for in the final analysis these Western religious creeds stem from Eastern wisdom. The Christ event is a different matter. For that is an actual event. It stands as a fact within the evolution of the earth. Yet the ways and means of understanding what came to pass through the Mystery of Golgotha were drawn during the first Christian centuries entirely from Eastern wisdom. It was through this wisdom that the fundamental event of Christianity was originally understood. Everything progresses, however. What had once been present in Eastern primeval wisdom—attained through Inspiration—spread from the East to Greece and is still recognizable as art. For Greek art was, to be sure, bound up with experiences different from those usually connected with art today. In Greek art one could still experience what Goethe strove to regain when he spoke of the deepest urge within him: he to whom nature begins to unveil her manifest secrets longs for her worthiest interpreter—art. For the Greeks, art was a way to slip into the secrets of world existence, a manifestation not merely of human fantasy but of what arises in the interaction between this facul-

97

ty and the revelations of the spiritual world revealed through Inspiration. That which still flowed through Greek art, however, became more and more diluted, until finally it became the content of the Western religious creeds. We thus must conceive the source of the primeval wisdom as fully substantial spiritual life that becomes impoverished as evolution proceeds and provides the content of religious creeds when it finally reaches the Western world. Human beings who are constitutionally suited for a later epoch therefore can find in this diluted form of spiritual life only something to be viewed with skepticism. And in the final analysis it is nothing other than the reaction of the Western temperament [*Gemüt*] to the now decadent Eastern wisdom that gradually produces atheistic skepticism in the West. This skepticism is bound to become more and more widespread unless it is countered with a different stream of spiritual life.

Just as little as a creature that has reached a certain stage of development—let us say has undergone a certain aging process—can be made young again in every respect, so little can a form of spiritual life be made young again when it has reached old age. The religious creeds of the West, which are descendants of the primeval wisdom of the East, can yield nothing that would fully satisfy Western humanity again when it advances beyond the knowledge provided during the past three or four centuries by science and observation of nature. An ever-more profound skepticism is bound to arise, and anyone who has insight into the processes of world evolution can say with assurance that a trend of development from East to West must necessarily lead to an increasingly pronounced skepticism when it is taken up by souls who are becoming more and more deeply imbued with the fruits of Western civilization. Skepticism is merely the march of the spiritual life from East to West, and it must be

98

countered with a different spiritual stream flowing henceforth from West to East. We ourselves are living at the crossing-point of these spiritual streams, and in the further course of these considerations we will want to see how this is so.

But first it must be emphasized that the Western temperament is constitutionally predisposed to follow a path of development leading to the higher worlds different from that of the Eastern temperament. Just as the Eastern temperament strives initially for Inspiration and possesses the racial qualities suitable for this, the Western temperament, because of its peculiar qualities (they are at present not so much racial qualities as qualities of soul) strives for Imagination. It is no longer the experience of the musical element in mantric aphorisms to which we as Westerners should aspire but something else. As Westerners we should strive in such a way that we do not pursue with particular vigor the path that opens out when the soul-spirit emerges from the physical body but rather the path that presents itself later, when the soul-spirit must again unite with the physical organism by consciously grasping the physical body. We see the natural manifestation of this in the emergence of the bodily instinct: whereas Eastern man sought his wisdom more by sublimating the forces at work between birth and the seventh year, Western man is better fitted to develop the forces at work between the time of the change of teeth and puberty, in that there is lifted up into the soul-spirit that which is natural for this epoch of humanity. We come to this when, just as in emerging from the body we carry the ego with us into the realm of Inspiration, we now leave the ego outside when we delve again into the body. We leave it outside, but not in idleness, not forgetting or surrendering it, not suppressing it into unconsciousness, but rather conjoining it with pure thinking, with clear, keen thinking, so that finally one has this inner experience: my ego is totally suffused with all the clear thinking of which I have become capable. One can experience just this delving down into the body in a very clear and distinct manner. And at this point you will perhaps allow me to relate a personal experience, because it will help you to understand what I really mean.

I have spoken to you about the conception underlying my book, *The Philosophy of Freedom*. This book is actually a modest attempt to win through to pure thinking, the pure thinking in which the ego can live and maintain a firm footing. Then, when pure thinking has been grasped in this way, one can strive for something else. This thinking, left in the power of an ego that now feels itself to be liberated within free spirituality [*frei und unabhängig in freier Geistigkeit*], can then be excluded from the process of perception. Whereas in ordinary life one sees color, let us say, and at the same time imbues the color with conceptual activity, one can now extract the concepts from the entire process of elaborating percepts and draw the percept itself directly into one's bodily constitution.

Goethe undertook to do this and has already taken the first steps in this direction. Read the last chapter of his *Theory of Colors*, entitled "The Sensory-Moral Effect of Color": in every color-effect he experiences something that unites itself profoundly not only with the faculty of perception but with the whole man. He experiences yellow and scarlet as "attacking" colors, penetrating him, as it were, through and through, filling him with warmth, while he regards blue and violet as colors that draw one out of oneself, as cold colors. The whole man experiences something in the act of sense perception. Sense perception, together with its content, passes down into the organism, and the ego with its pure thought content remains, so to speak, hovering above. We exclude thinking inasmuch as we take into and fill ourselves with the whole content of the

100

perception, instead of weakening it with concepts, as we usually do. We train ourselves specially to achieve this by systematically pursuing what came to be practiced in a decadent form by the men of the East. Instead of grasping the content of the perception in pure, strictly logical thought, we grasp it symbolically, in pictures, allowing it to stream into us as a result of a kind of detour around thinking. We steep ourselves in the richness of the colors, the richness of the tone, by learning to experience the images inwardly, not in terms of thought but as pictures, as symbols. Because we do not suffuse our inner life with the thought content, as the psychology of association would have it, but with the content of perception indicated through symbols and pictures, the living inner forces of the etheric and astral bodies stream toward us from within, and we come to know the depths of consciousness and of the soul. It is in this way that genuine knowledge of the inner nature of man is acquired, and not by means of the blathering mysticism that nebulous minds often claim to be a way to the God within. This mysticism leads to nothing but abstraction and cannot satisfy anyone who wishes to become a man in the full sense of the word.

If one desires to do real research concerning human physiology, thinking must be excluded and the picture-forming activity sent inward, so that the physical organism reacts by creating Imaginations. This is a path that is only just beginning in the development of Western culture, but it is the path that must be trodden if the influence that streams over from the East, and would lead to decadence if it alone were to prevail, is to be confronted with something capable of opposing it, so that our civilization may take a path of ascent and not of decline. Generally speaking, however, it can be said that human language itself is not yet sufficiently developed to be able to give full expression to the experiences that one undergoes in the inner recesses of the soul. And it is at this point that I would like to relate a personal experience to you.

Many years ago, in a different context, I made an attempt to give expression to what might be called a science of the human senses. In spoken lectures I succeeded to some extent in putting this science of the twelve senses into words, because in speaking it is more possible to turn language this way and that and ensure understanding by means of repetitions, so that the deficiencies of our language, which is not yet capable of expressing these supersensible things, is not so strongly felt. Strangely enough, however, when I wanted many years ago to write down what I had given as actual anthroposophy in order to put it into a form suitable for a book, the outer experiences on being interiorized became so sensitive that language simply failed to provide the words, and I believe that the beginning of the text —several sheets of print—lay for some five or six years at the printer's. It was because I wanted to write the whole book in the style in which it began that I could not continue writing, for the simple reason that at the stage of development I had then reached, language refused to furnish the means for what I wished to achieve. Afterward I became overloaded with work, and I still have not been able to finish the book. Anyone who is less conscientious about what he communicates to his fellow men out of the spiritual world might perhaps smile at the idea of being held up in this way by a temporarily insurmountable difficulty. But whoever really experiences and can permeate with a full sense of responsibility what occurs when one attempts to describe the path that Western humanity must follow to attain Imagination knows that to find the right words entails a great deal of effort. As a meditative schooling it is relatively easy to describe, and this has been done in my book, *Knowledge of the Higher Worlds and its Attainment.* If one's aim, however, is to achieve definite results such as that of describing the essential nature of man's senses—a part, therefore, of the inner makeup and constitution of humanity—it is then that one encounters the difficulty of grasping Imaginations and presenting them in sharp contours by means of words.

Nevertheless, this is the path that Western humanity must follow. And just as the man of the East was able to experience through his mantras the entry into the spiritual nature of the external world, so must the Westerner, leaving aside the entire psychology of association, learn to enter into his own being by attaining the realm of Imagination. Only by penetrating into the realm of Imagination will he acquire the true knowledge of humanity that is necessary in order for humanity to progress. And because we in the West must live much more consciously than the men of the East, we cannot simply say: whether or not humanity will gradually attain this realm of Imagination is something that can be left to the future. No—this world of Imagination, because we have passed into the stage of conscious human evolution, must be striven for consciously; there can be no halting at certain stages. For what happens if one halts at a certain stage? Then one does not meet the ever-increasing spread of skepticism from East to West with the right countermeasures but with measures that result from the soul-spirit uniting too radically, too deeply and unconsciously, with the physical body, so that too strong a connection is formed between the soul-spirit and the physical body.

Yes, it is indeed possible for a human being not only to think materialistically but to *be* a materialist, because the soul-spirit is too strongly linked with the physical body. In such a man the ego does not live freely in the concepts of pure thinking he has attained. If one descends into the body with pictorial perception, one delves with the ego and the concepts into the body. And if one then spreads this around and suffuses it throughout humanity, it gives rise to a spiritual phenomenon well known to us—dogmatism of all kinds. Dogmatism is nothing other than the translation into the realm of the soul-spirit of a condition that at a lower stage manifests itself pathologically as agoraphobia and the like, and that—because these things are related—also shows itself in something else, which is a metamorphosis of fear, in superstition of every variety. An unconscious urge toward Imagination is held back through powerful agencies, and this gives rise to dogmatism of all types. These types of dogmatism must gradually be replaced by what is achieved when the world of ideas is kept within the sphere of the ego; when progress is made toward Imagination, the true nature of man is experienced inwardly, and this Western path into the spiritual world is followed in a different way. It is this other path through Imagination that must establish the stream of spiritual science, the process of spiritual evolution that must make its way from West to East if humanity is to progress. It is supremely important at the present time, however, for humanity to recognize what the true path of Imagination should be, what path must be taken by Western spiritual science if it is to be a match for the Inspiration and its fruits that were attained by ancient Eastern wisdom in a form suited to the racial characteristics of those peoples. Only if we are able to confront the now decadent Inspiration of the East with Imaginations which, sustained by the spirit and saturated with reality, have arisen along the path leading to a higher spiritual culture; only if we can call this culture into existence as a stream of spiritual life flowing from West to East, are we bringing to fulfillment what is actually living deep within the impulses for which humanity is striving. It is these impulses that are now exploding in social cataclysms because they cannot find other expression.

In tomorrow's lecture we will speak further of the path of Imagination and of how the way to the higher worlds is envisaged by anthroposophical spiritual science.

VIII

Yesterday I attempted to show the methods employed by Eastern spirituality for approaching the spiritual world and pointed out how anybody who wished to pursue this path into the supersensible more or less dispensed with the bridge linking him with his fellow men. He chose a path different from that which establishes communication within society by means of language, thought, and perception of the ego. And I showed how it was initially attempted not to understand through the word what one's fellow man wished to say, what one wants to understand from him, but to live within the words. This process of living within the word was enhanced by forming the words into certain aphorisms. One lived in these and repeated them, so that the forces accrued in the soul by this process were strengthened further by repetition. And I showed how something was achieved in the condition of the soul that might be called a state of Inspiration, in the sense in which I have used the word, except that the sages of the ancient East were, of course, members of their race: their ego-consciousness was much less developed than in later epochs of human evolution. They thus entered into the spiritual world in a more instinctive manner, and because the whole thing was instinctive and thus resulted, in a sense, from a healthy drive within human nature, in the earliest times it could not lead to the pathological afflictions of which we have also spoken. In later times steps were taken by the so-called Mysteries to guard against the rise of such afflictions as I have described to you. I said

105

that those Westerners who desire to gain knowledge of the spiritual world must approach this in another way. Humanity has progressed in the interim. Different soul faculties have evolved, so that one cannot simply renew the ancient Eastern path of spiritual development. Within the realm of spiritual life one cannot long to return in a reactionary manner to prehistoric or earlier historical periods of human evolution. For Western civilization, the path leading into the spiritual worlds is that of Imagination. This faculty of Imagination, however, must be integrated organically into the life of the soul as a whole. This can come about in the most varied ways, just as the Eastern path of development was not unequivocally predetermined but could take numerous different courses. Today I would like to describe the path into the spiritual world that conforms to the needs of Western civilization and is particularly suited to anyone immersed in the scientific life of the West.

In my book, *Knowledge of the Higher Worlds and its Attainment*, I have described an entirely safe path leading to the supersensible, but I describe it in such a way that it applies for everybody, above all for those who have not devoted their lives to science. Today I shall describe a path into the supersensible that is much more for the scientist. All my experience has taught me that for such a scientist a kind of precondition for this cognitional striving is to take up what is presented in my book, *The Philosophy of Freedom*. I will explain what I mean by this. This book, *The Philosophy of Freedom*, was not written with the same intent as most books written today. Nowadays books are written simply in order to inform the reader of the book's subject-matter, so that the reader learns the book's contents in accordance with his education, his scientific training, or the special knowledge he already possesses. This was not my primary intention in writing *The Philosophy of Freedom*, and thus it will

106

not be popular with those who read books only to acquire information. The purpose of the book is to make the reader directly engage his thinking activity on every page.

In a sense, the book is only a kind of musical score that one must read with inner thought activity in order to progress, as the result of one's own efforts, from one thought to the next. The book constantly presupposes the mental collaboration of the reader. Moreover, the book presupposes that which the soul becomes in the process of such mental exertion. Anyone who has really worked through this book with his own inner thinking activity and cannot confess that he has come to know himself in a part of his inner life in which he had not known himself previously has not read *The Philosophy of Freedom* properly. One should feel that one is being lifted out of one's usual thinking [*Vorstellen*] into a thinking independent of the senses [*ein sinnlichkeitsfreies Denken*], in which one is fully immersed, so that one feels free of the conditions of physical existence. Whoever cannot confess this to himself has actually misunderstood the book. One should be able to say to oneself: now I know, as a result of the inner thought activity I myself have expended, what pure thinking actually is.

The strange thing is that most Western philosophers totally deny the reality of the very thing that my *Philosophy of Freedom* seeks to awaken as something real in the soul of the reader. Countless philosophers have expounded the view that pure thinking does not exist but is bound to contain traces, however diluted, of sense perception. A strong impression is left that philosophers who maintain this have never really studied mathematics or gone into the difference between analytical and empirical mechanics. Specialization, however, has already grown to such an extent that nowadays philosophy is often pursued by people totally lacking any knowledge of mathematical thinking. The pursuit of philos-

107

ophy is actually impossible without a grasp of at least the spirit of mathematical thinking. We have seen what Goethe's attitude was toward this spirit of mathematical thinking, even though he made no claim himself to any special training in mathematics. Many thus would deny the existence of the very faculty I would like those who study *The Philosophy of Freedom* to acquire.

And now let us imagine a reader who simply sets about working through *The Philosophy of Freedom* within the context of his ordinary consciousness in the way I have described: he will, of course, not be able to claim that he has been transported into a supersensible world. For I intentionally wrote *The Philosophy of Freedom* in the way that I did so that it would present itself to the world initially as a purely philosophical work. Just think what a disservice would have been accorded anthroposophically oriented spiritual science if I had begun immediately with spiritual scientific writings! These writings would, of course, have been disregarded by all trained philosophers as the worst kind of dilettantism, as the efforts of an amateur. To begin with I had to write purely philosophically. I had to present the world with something thought out philosophically in the strict sense, though it transcended the normal bounds of philosophy. At some point, however, the transition had to be made from a merely philosophical and scientific kind of writing to a spiritual scientific writing. This occurred at a time when I was invited to write a special chapter about Goethe's scientific writings for a German biography of Goethe. This was at the end of the last century, in the 1890s. And so I was to write the chapter on Goethe's scientific writings: I had, in fact, finished it and sent it to the publisher when there appeared another work of mine, called *Mysticism at the Dawn of the Modern Age*. The book was a bridge between pure philosophy and an anthroposophical orientation. When this work came out, my manuscript was returned to me by the publisher, who had enclosed nothing but my fee so that I would not make a fuss, for thereby the legal obligations had been met. Among the learned pedants, there was obviously no interest in anything—not even a single chapter devoted to the development of Goethe's attitude toward natural science—written by one who had authored this book on mysticism.

I will now assume that *The Philosophy of Freedom* has been worked through already with one's ordinary consciousness in the way described. Now we are in the right frame of mind for our souls to undertake in a healthy way what I described yesterday, if only very briefly, as the path leading into Imagination. It is possible to pursue this path in a way consonant with Western life if we attempt to surrender ourselves completely to the world of outer phenomena, so that we allow them to work upon us without thinking about them but still perceiving them. In ordinary waking life, you will agree, we are constantly perceiving, but actually in the very process of doing so we are continually saturating our percepts with concepts; in scientific thinking we interweave percepts and concepts entirely systematically, building up systems of concepts and so on. By having acquired the capacity for the kind of thinking that gradually emerges from *The Philosophy of Freedom*, one can become capable of such acute inner activity that one can exclude and suppress conceptual thinking from the process of perception and surrender oneself to bare percepts. But there is something else we can do in order to strengthen the forces of the soul and absorb percepts unelaborated by concepts. One can, moreover, refrain from formulating the judgments that arise when these percepts are joined to concepts and create instead symbolic images, or images of another sort, alongside the images seen by the eye, heard by the ear, and rendered by the senses of warmth, touch, and so on. If we thus bring our activity of perception into a state of flux, infusing it with life and movement, not as we do when forming concepts but by elaborating perception symbolically or artistically, we will develop much sooner the power of allowing the percepts to permeate us as such. An excellent preparation for this kind of cognition is to school oneself rigorously in what I have characterized as phenomenalism, as elaboration of phenomena. If one has really striven not to allow inertia to carry one through the veil of sense perception upon reaching the boundary of the material world, in order to look for all kinds of metaphysical explanations in terms of atoms and molecules, but has instead used concepts to set the phenomena in order and follow them through to the archetypal phenomena, one has already undergone a training that enables one to isolate the phenomena from everything conceptual. And if one still symbolizes the phenomena, turns them into images, one acquires a potent soul force enabling one to absorb the external world free from concepts.

Obviously we cannot expect to achieve this quickly. Spiritual research demands of us far more than research in a laboratory or observatory. It demands above all an intense effort of the individual will. If one has practiced such an inner representation of symbolic images for a certain length of time and striven in addition to dwell contemplatively upon images that one keeps present in the soul in a way analogous to the mental representation of phenomena, images that otherwise only pass away when we race from sensation to sensation, from experience to experience; if one has accustomed oneself to dwell contemplatively for longer and longer periods of time upon an image that one has fully understood, that one has formed oneself or taken at somebody else's suggestion so that it cannot be a reminiscence, and if one repeats this process again and again, one strengthens one's inner soul forces and finally realizes that one experiences something of which one previously had no inkling. The only way to obtain even an approximate idea of such an experience, which takes place only in one's inner being—one must be very careful not to misunderstand this —is to recall particularly lively dream-images. One must keep in mind, however, that dream-images are always reminiscences that can never be related directly to anything external and are thus a sort of reaction coming toward one out of one's own inner self. If one experiences to the full the images formed in the way described above, this is something entirely real, and one begins to understand that one is encountering within oneself the spiritual element that actuates the processes of growth, that is the power of growth. One realizes that one has entered into a part of one's human constitution, something within one; something that unites itself with one; something that is active within but that one previously had experienced only unconsciously. Experienced unconsciously in what way?

I have told you that from birth until the change of teeth a soul-spiritual entity is at work structuring the human being and that this then emancipates itself to an extent. Later, between the change of teeth and puberty, another such soul-spiritual entity, which dips down in a way into the physical body, awakens the erotic drives and much else as well. All this occurs unconsciously. If, however, we use fully consciously such measures of soul as I have described to observe this permeation of the physical organism by the soul-spiritual, one sees how such processes work within man and how man is actually given over to the external world continually, from birth onward. Nowadays this giving-over of oneself to the external world is held to be nothing but abstract perception or abstract cognition. This is not so. We are surrounded by a world of color, sound, and warmth and by all kinds of sense impressions. By elaborating these with our concepts we create yet further impressions that have an effect on us. By experiencing all this consciously we come to see that in the unconscious experience of color- and sound-impressions that we have from childhood onward there is something spiritual that suffuses our organization. And when, for example, we take up the sense of love between the change of teeth and puberty, this is not something originating in the physical body but rather something that the cosmos gives us through the colors, sounds, and streaming warmth that reach us. Warmth is something other than warmth; light something other than light in the physical sense; sound is something other than physical sound. Through our sense impressions we are conscious only of what I would term external sound and external color. And when we surrender ourselves to nature, we do not encounter the ether-waves, atoms, and so on of which modern physics and physiology dream; rather, it is spiritual forces that are at work, forces that fashion us between birth and death into what we are as human beings. Once we tread the path of knowledge I have described, we become aware that it is the external world that forms us. We become best able to observe consciously what lives and embodies itself within us when we acquire above all a clear sense that spirit is at work in the external world. It is of all things phenomenology that enables us to perceive how spirit works within the external world. It is through phenomenology, and not abstract metaphysics, that we attain knowledge of the spirit by consciously observing, by raising to consciousness, what otherwise we would do unconsciously, by observing how, through the sense world, spiritual forces enter our being and work formatively upon it.

Yesterday I pointed out to you that the Eastern sage in a way disregards the significance of speech, thought, and the perception of the ego. He experiences these things differently and cultivates a different attitude of soul toward these things, because language, perception of thoughts, and perception of the ego initially tend to lead us away from the spiritual world into social contact with other human beings. In everyday physical existence we purchase our social life at the price of listening right through language, looking through thoughts, and feeling our way right through the perception of the ego. The Eastern sage took upon himself not to listen right through the word but to live within it. He took upon himself not to look right through the thought but to live within the thought, and so forth. We in the West have as our task more to contemplate man himself in following the path into supersensible worlds.

At this point it must be remembered that man bears a certain kind of sensory organization within as well. I have already described the three inner senses through which he becomes aware of his inner being, just as he perceives what goes on outside him. We have a sense of balance by means of which we sense the spatial orientation appropriate to us as human beings and are thereby able to work inside it with our will. We have a sense of movement by means of which we know that we are moving even in the dark: we know this from an inner sensing and not merely because we perceive our changing relationship to other objects we pass. We have an actual inner sense of movement. And we have a sense of life, by means of which we can perceive our general state of well-being, the constant changes in the inner condition of our life forces. These three inner senses work together with the will during man's first seven years. We are guided by our sense of balance, and a being who initially cannot move at all and later can only crawl is transformed into one who can stand upright and walk. This ability to walk upright is effected by the sense of balance, which places us into the world. The sense of movement and the sense of life likewise

113

contribute toward the development of our full humanity. Anybody who is capable of applying the standards of objective observation employed in the scientist's laboratory to the development of man's physical body and his soul-spirit will soon discover how the forces that worked formatively upon man principally during the first seven years emancipate themselves and begin to assume a different aspect from the time of the change of teeth onward. By this time a person is less intensively connected to that within than he was as a child. A child is closely bound up inwardly with human equilibrium, movement, and life. Something else, however, is evolving simultaneously during this emancipation of balance, movement, and life. There takes place a certain adjustment of the three other senses: the senses of smell, taste, and touch. It is extremely interesting to observe in detail the way in which a child gradually finds his way into life, orienting himself by means of the senses of taste, smell, and touch. Of course, this can be seen most obviously in early life, but anybody trained to do so can see it clearly enough later on as well. In a certain way, the child pushes out of himself balance, movement, and life but at the same time draws more into himself the qualities of the sense of smell, the sense of taste, and the sense of touch. In the course of an extended phase of development the one is, so to speak, exhaled and the other inhaled, so that the forces of balance, movement, and life, which press from within outward, and the qualitative orientations of smell, taste, and touch, which press from without inward, meet within our organism. This is effected by the interpenetration of the two sense-triads. As a result of this interpenetration, there arises within man a firm sense of self; in this way man first experiences himself as a true ego. Now we are cut off from the spirituality of the external world by speech and by our faculties of perceiving thoughts and perceiving the egos of others—and rightly so,

114

for if it were otherwise we could never in this physical life become social beings—in just the same way, inasmuch as the qualities of smell, taste, and touch encounter balance, movement, and life, we are inwardly cut off from the triad life, movement, and balance, which would otherwise reveal itself to us directly. The experiences of the senses of smell, taste, and touch place themselves, as it were, in front of what we would otherwise experience through our sense of balance, our sense of movement, and our sense of life. And the result of this development toward Imagination of which I have spoken consists in this: the Oriental comes to a halt at language in order to live within it; he halts at the thought in order to live there; he halts at the perception of the ego in order to live within it. By these means he makes his way outward into the spiritual world. The Oriental comes to a halt within these; we, by striving for Imagination, by a kind of absorption of external percepts devoid of concepts, engage in an activity that is in a way the opposite of that in which the Oriental engages with regard to language, perception of thoughts, and perception of the ego. The Oriental comes to a halt at these and enters into them. In striving for Imagination, however, one wends one's way through the sensations of smell, taste, and touch, penetrating into the inner realm so that, by one's remaining undisturbed by sensations of smell, taste, and touch, the experiences stemming from balance, movement, and life come forth to meet one.

It is a great moment when one has penetrated through what I have described as the sense-triad of taste, smell, and touch, and one confronts the naked essence of movement, balance, and life.

With such a preparation behind us, it is interesting to study what Western mysticism often sets forth. Most certainly, I am very far from decrying the elements of poetry, beauty, and imaginative expression in the writings of many

115

mystics. I most certainly admire what, for instance, St. Theresa, Mechthild of Magdeburg, and others have to tell us, and indeed Meister Eckhart and Johannes Tauler. But all that arises in this way reveals itself to the true spiritual scientist as something that arises when one traverses the inward-leading path yet does not penetrate beyond the region of smell, taste, and touch. Read what has been written by individuals who have described with particular clarity what they have experienced in this way. They speak of a tasting of that within, of a tasting regarding what exists as soul-spirit in man's inner being; they also speak of a smelling and, in a certain sense, of a touching. And anybody who knows how to read Mechthild of Magdeburg, for instance, or St. Theresa, in the right way will see that they follow this inward path but never penetrate right through taste, smell, and touch. They use beautiful poetic imagery for their descriptions, but they are speaking only of how one can touch, savor, and sniff oneself inwardly.

For it is far less agreeable to see the true nature of reality with senses that are developed truly spiritually than to read the accounts given by voluptuous mysticism—the only term for it—which in the final analysis only gratifies a refined, inward-looking egotism of soul. As I say, much as this mysticism is to be admired—and I do admire it—the true spiritual scientist must realize that it stops halfway: what is manifest in the splendid poetic imagery of Mechthild of Magdeburg, St. Theresa, and the others is really only what is smelt, tasted, and touched before breaking through into the actual inner realm. Truth is occasionally unpleasant, and at times perhaps even cruel, but modern humanity has no business becoming rickety in soul by following a nebulous, imperfect mysticism. What is required today is to penetrate into man's true inner nature with strength of spirit, with the same strength we have achieved in a much

116

more disciplined way for the external world by pursuing natural science. And it is not in vain that we have achieved this. Natural science must not be undervalued! Indeed, we must seek to acquire the disciplined and methodical side of natural science. And it is precisely when one has assimilated this scientific method that one appreciates the achievements of a nebulous mysticism at their true worth, but one also knows that this nebulous mysticism is not what spiritual science must foster. On the contrary, the task of spiritual science is to seek clear comprehension of man's own inner being, whereby a clear, spiritual understanding of the external world is made possible in turn.

I know that if I did not speak in the way that truth demands I could enjoy the support of every nebulous, blathering mystic who takes up mysticism in order to satisfy his voluptuous soul. That cannot be our concern here, however; rather, we must seek forces that can be used for life, spiritual forces that are capable of informing our scientific and social life.

When one has penetrated as far as that which lives in the sense of balance, the sense of life, and the sense of movement, one has reached something that one experiences initially as the true inner being of man because of its transparency. The very nature of the thing shows us that we cannot penetrate any deeper. But then again one has more than enough at this initial stage, for what we discover is not the stuff of nebulous, mystical dreams. What one finds is a true organology, and above all one finds within oneself the essence of that which is within equilibrium, of that which is in movement, of that which is suffused with life. One finds this within oneself.

Then, after experiencing this, something entirely extraordinary has occurred. Then, at the appropriate moment, one begins to notice something. An essential prerequisite is, as I have said, to have thought through *The Philosophy of Freedom* beforehand. This is then left, so to speak, to one side, while pursuing the inner path of contemplation, of meditation. One has advanced as far as balance, movement, and life. One lives within this life, this movement, this balance. Entirely parallel with our pursuit of the way of contemplation and meditation but without any other activity on our part, our thinking regarding *The Philosophy of Freedom* has undergone a transformation. What can be experienced in such a philosophy of freedom in pure thinking has, as a result of our having worked inwardly on our souls in another sphere, become something utterly different. It has become fuller, richer in content. While on the one hand we have penetrated into our inner being and have deepened our power of Imagination, on the other hand we have raised what resulted from our mental work on *The Philosophy of Freedom* up out of ordinary consciousness. Thoughts that formerly had floated more or less abstractly within pure thinking have been transformed into substantial forces that are alive in our consciousness: what once was pure thought is now Inspiration. We have developed Imagination, and pure thinking has become Inspiration. Following this path further, we become able to keep apart what we have gained following two paths that must be sharply differentiated: on the one hand, what we have obtained as Inspiration from pure thinking—the life that at a lower level is thinking, and then becomes a thinking raised to Inspiration—and on the other hand what we experience as conditions of equilibrium, movement, and life. Now we can bring these modes of experience together. We can unite the inner with the outer. The fusion of Imagination and Inspiration brings us in turn to Intuition. What have we accomplished now? Well, I would like to answer this question by approaching it from another side. First of all I must draw attention to the steps taken by the Oriental who wishes to rise further after having schooled himself by means of the mantras, after having lived within the language, within the word. He now learns not only to live in the rhythms of language but also in a certain way to experience breathing consciously, in a certain way to experience breathing artificially by altering it in the most varied ways. For him this is the next highest step—but again not something that can be taken over directly by the West. What does the Eastern student of yoga attain by surrendering himself to conscious, regulated, varied breathing? Oh, he experiences something quite extraordinary when he inhales. When inhaling he experiences a quality of air that is not found when we experience air as a purely physical substance but only when we unite ourselves with the air and thus comprehend it spiritually. As he breathes in, a genuine student of yoga experiences something that works formatively upon his whole being, that works spiritually; something that does not expend itself in the life between birth and death, but, entering into us through the spirituality of the outer air, engenders in us something that passes with us through the portal of death. To experience the breathing process consciously means taking part in something that persists when we have laid aside the physical body. For to experience the breathing process consciously is to experience the reaction of our inner being to inhalation. In experiencing this we experience something that preceded birth in our existence as soul-spirit—or let us say preceded our conception—something that had already cooperated in shaping us as embryos and then continued to work within our organism in childhood. To grasp the breathing process consciously means to comprehend ourselves beyond birth and death. The advance from an experience of the aphorism and the word to an experience of the breathing process represented a further penetration into an inspired comprehension of the eternal in man. We Westerners must experience much the same thing—but in a different sphere.

What, in fact, is the process of perception? It is nothing but a modified process of inhalation. As we breathe in, the air presses upon our diaphragm and upon the whole of our being. Cerebral fluid is forced up through the spinal column into the brain. In this way a connection is established between breathing and cerebral activity. And the part of the breathing that can be discerned as active within the brain works upon our sense activity as perception. Perception is thus a kind of branch of inhalation. In exhalation, on the other hand, cerebral fluid descends and exerts pressure on the circulation of the blood. The descent of cerebral fluid is bound up with the activity of the will and also of exhalation. Anybody who really studies *The Philosophy of Freedom*, however, will discover that when we achieve pure thinking, thinking and willing coincide. Pure thinking is fundamentally an expression of will. Thus pure thinking turns out to be related to what the Oriental experienced in the process of exhalation. Pure thinking is related to exhalation just as perception is related to inhalation. We have to go through the same process as the yogi but in a way that is, so to speak, pushed back more into the inner life. Yoga depends upon a regulation of the breathing, both inhalation and exhalation, and in this way comes into contact with the eternal in man. What can Western man do? He can raise into clear soul experiences perception on the one hand and thinking on the other. He can unite in his inner experience perception and thinking, which are otherwise united only abstractly, formally, and passively, so that inwardly, in his soul-spirit, he has the same experience as he has physically in breathing in and out. Inhalation and exhalation are physical experiences: when they are harmonized, one consciously experiences the eternal. In everyday life we experience thinking and perception. By bringing mobility into the life of the soul, one experiences the pendulum, the rhythm, the continual interpenetrating vibration of perception and thinking. A higher reality evolves for the Oriental in the process of inhalation and exhalation; the Westerner achieves a kind of breathing of the soul-spirit in place of the physical breathing of the yogi. He achieves this by developing within himself the living process of modified inhalation in perception and modified exhalation in pure thinking, by weaving together concept, thinking, and perceiving. And gradually, by means of this rhythmic pulse, by means of this rhythmic breathing process in perception and thinking, he struggles to rise up to spiritual reality in Imagination, Inspiration, and Intuition. And when I indicated in my book *The Philosophy of Freedom*, at first only philosophically, that reality arises out of the interpenetration of perception and thinking, I intended, because the book was meant as a schooling for the soul, to show what Western man can do in order to enter the spiritual world itself. The Oriental says: systole, diastole; inhalation, exhalation. In place of these the Westerner must put perception and thinking. Where the Oriental speaks of the development of physical breathing, we in the West say: development of a breathing of the soul-spirit within the cognitional process through perception and thinking.

All this had to be contrasted with what can be experienced as a kind of dead end in Western spiritual evolution. Let me explain what I mean. In 1841 Michelet, the Berlin philosopher, published posthumously Hegel's works on natural philosophy. Hegel had worked at the end of the eighteenth century, together with Schelling, at laying the foundations of a system of natural philosophy. Schelling, as a young firebrand, had constructed his natural philosophy in a remarkable way out of what he called "intellectual intuition" [*intellektuale Anschauung*]. He reached a point, however, where he could make no further progress. He immersed himself in the mystics at a certain point. His work, *Bruno, or Concerning the Divine and Natural Principle in Things*, and his fine treatise on human freedom and the origin of evil testify so wonderfully to this immersion. But for all this he could make no progress and began to hold back from expressing himself at all. He kept promising to follow up with a philosophy that would reveal the true nature of those hidden forces at which his earlier natural philosophy had only hinted. When Michelet published Hegel's natural philosophy in 1841, Schelling's long-expected and oft-promised "philosophy of revelation" had still not been vouchsafed to the public. He was summoned to Berlin. What he had to offer, however, was not the actual spirit that was to permeate the natural philosophy he had founded. He had striven for an intellectual intuition. He ground to a halt at this point, because he was unable to use Imagination to enter the sphere of which I spoke to you today. And so he was stuck there. Hegel, who had a more rational intellect, had taken over Schelling's thoughts and carried them further by applying pure thinking to the observation of nature. That was the origin of Hegel's natural philosophy. And so one had Schelling's unfulfilled promise to bring forth nature out of the spirit, and then one had Hegel's natural philosophy, which was discarded by science in the second half of the nineteenth century. It was misunderstood, to be sure, but it was bound to remain so, because it was impossible to gain any kind of connection to the ideas contained in Hegel's natural philosophy with regard to phenomenology, the true observation of nature. It is a kind of wonderful incident: Schelling traveling from Munich to Berlin, where great things are expected of him, and it turns out that he has nothing to say. It was a disappointment for all who believed that through Hegel's natural

122

philosophy revelations about nature would emerge from pure thinking. Thus it was in a way demonstrated historically, in that Schelling had attained the level of intellectual intuition but not that of genuine Imagination and in that Hegel showed as well that if pure thinking does not lead on to Imagination or to Inspiration—that is, to the level of nature's secrets . . . it was shown that the evolution of the West had thereby run up against a dead end. There was as yet nothing to counter what had come over from the Orient and engendered skepticism; one could counter with nothing that was suffused with the spirit. And anyone who had immersed himself lovingly in Schelling and Hegel and has thus been able to see, with love in his heart, the limitations of Western philosophy, had to strive for anthroposophy. He had to strive to bring about an anthroposophically oriented spiritual science for the West, so that we will possess something that works creatively in the spirit, just as the East had worked in the spirit through systole and diastole in their interaction. We in the West can allow perception and thinking to resound through one another in the soul-spirit [das geistig-seelische Ineinanderklingenlassen], through which we can rise to something more than a merely abstract science. It opens the way to a living science, which is the only kind of science that enables us to dwell within the element of truth. After all the failures of the Kantian, Schellingian, and Hegelian philosophies, we need a philosophy that, by revealing the way of the spirit, can show the real relationship between truth and science, a spiritualized science, in which truth can really live to the great benefit of future human evolution.

Translators' Notes

1. ". . . that social renewal must begin with the renewal of our thinking." The original German (". . . *dass die soziale Erneuerung vom Geiste ausgehen musse*") might be translated alternately "that social renewal must proceed from the spirit." The German word "*Geist*"—bane of all who would translate German into English—embraces two meanings that remain in English quite distinct: "mind" and "spirit." The translator must choose one, even though the German always implies both. If he chooses the former, he runs the risk of seriously distorting the author's intentions (as did the man who translated Hegel's *Phanomenologie des Geistes* as *The Phenomenology of Mind*). If he chooses the latter, he flies in the face of the dubious connotations that "spirit" and "spiritual" convey—no doubt as a result of the basically empirical cast of English thought. Although "*Geist*" as Steiner uses it should almost invariably be translated "spirit" (which of course comprehends "mind"), here the context has led us to choose the more restricted meaning.

2. *Wie erlangt man Erkenntnisse der hoheren Welten?*, Berlin, 1909 (*Knowledge of the Higher Worlds and its Attainment*. Anthroposophic Press, Spring Valley, N.Y., reprinted 1983).

3. *Die Philosophie der Freiheit*, Berlin, 1894 (*The Philosophy of Freedom*. Anthroposophic Press, Spring Valley, N.Y., 1964). Earlier translations of this book (1922, 1938, and 1963) bore the title *The Philosophy of Spiritual Activity*, following a suggestion given by Steiner himself. The English word "freedom" connotes a passive state; the German "*Freiheit*" (as is clear from the following lecture), an objective basis for moral action achieved through intense inner activity.

4. *Grundlinien einer Erkenntnistheorie der Goetheschen Weltanschauung*, Berlin and Stuttgart, 1886 (*A Theory of Knowledge Based on Goethe's World Conception*. Anthroposophic Press, Spring Valley, N.Y.,1968).

5. *Die Rätsel der Philosophie in ihrer Geschichte als Umriss dargestellt*, Berlin, 1914 (*The Riddles of Philosophy*. Anthroposophic Press, Spring Valley, N.Y, 1973).

6. "Astraphobia" = morbid dread of storms; "agoraphobia" = morbid dread of crossing, or being in, open spaces.

7. *Theosophie. Einfuhrung in ubersinnliche Welterkenntnis und Menschenbestimmung*, Berlin, 1904 (*Theosophy. An Introduction to the Supersensible Knowledge of the World and the Destiny of Man*. Anthroposophic Press, Spring Valley, N.Y., 1971).

8. The German edition gives "claustrophobia" here, which seems to be a mistake.

125